Electrical installation calculations

Volume 1

CENTRE	Newark
CHECKED	S-W
ZONE	Grey
ZONE MARK / SUFFIX	621.3282 WAT
LOAN PERIOD	1 month

Electrical installation calculations

VOLUME I

SEVENTH EDITION

A. J. Watkins

Chris Kitcher

ELSEVIER

AMSTERDAM • BOSTON • HEIDELBERG • LONDON • NEW YORK
OXFORD • PARIS • SAN DIEGO • SAN FRANCISCO
SINGAPORE • SYDNEY • TOKYO
Newnes is an imprint of Elsevier

Newnes

Newnes is an imprint of Elsevier
Linacre House, Jordan Hill, Oxford OX2 8DP, UK
30 Corporate Drive, Suite 400, Burlington, MA 01803, USA

First edition 1957
Sixth edition 1988
Reprinted 2001, 2002, 2003 (twice), 2004
Seventh edition 2006
Reprinted 2006, 2007

British Library Cataloguing in Publication Data
A catalogue record for this book is available from the British Library

Library of Congress Cataloging-in-Publication Data
A catalog record for this book is available from the Library of Congress

ISBN–13: 978-0-7506-6782-1
ISBN–10: 0-7506-6782-6

For information on all Newnes publications
visit our website at www.newnespress.com

Printed and bound in *Great Britain*

07 08 09 10 10 9 8 7 6 5 4 3

Working together to grow
libraries in developing countries

www.elsevier.com | www.bookaid.org | www.sabre.org

ELSEVIER   BOOK AID International   Sabre Foundation

1081883
£9.99

Contents

Preface

Mathematics forms the essential foundation of electrical installation work. Without applying mathematical functions we would be unable to work out the size of a room which needs lighting or heating, the size and/or the number of the lights or heaters themselves, the number and/or the strength of the fixings required, or the size of the cables supplying them. We would be unable to accurately establish the rating of the fuse or circuit breaker needed to protect the circuits, or predict the necessary test results when testing the installation. Like it or not you will need to be able to carry out mathematics if you want to be an efficient and skilled electrician.

This book will show you how to perform the maths you will need to be a proficient electrician. It concentrates on the electronic calculator methods you would use in class and in the workplace. The book does not require you to have a deep understanding of how the mathematical calculations are performed – you are taken through each topic step by step, then you are given the opportunity yourself to carry out exercises at the end of each chapter. Throughout the book useful references are made to Amendment 2: 2004 BS 7671 Requirements for Electrical Regulations and the *IEE On-Site Guide*.

Simple cable selection methods are covered comprehensively in this volume so as to make it a useful tool for tradesmen involved in Part P of the building regulations, with more advanced calculations being added in Volume 2.

Volume 1 Electrical Installation Calculations originally written by A.J. Watkins and R.K. Parton has been the preferred book for many students looking to improve their mathematical understanding of the subject for many years. This edition has been newly updated not only to include modern methods, but also to cover all aspects of the new City and Guilds 2330 Certificate in Electrotechnical Technology.

Chris Kitcher

Use of calculators

Throughout Books 1 and 2, the use of a calculator is encouraged. Your calculator is a tool, and like any tool practice is required to perfect its use. A scientific calculator will be required, and although they differ in the way the functions are carried out the end result is the same.

The examples are given using a Casio fx-83MS. The figures printed on the button is the function performed when the button is pressed. To use the function in small letters above any button the *shift* button must be used.

PRACTICE IS IMPORTANT

Syntax error — Appears when the figures are entered in the wrong order.

x^2 — Multiplies a number by itself, i.e. $6 \times 6 = 36$. On the calculator this would be $6\ x^2 = 36$. When a number is multiplied by itself it is said to be *squared*.

x^3 — Multiplies a number by itself and then the total by itself again, i.e. when we enter 4 on calculator $x^3 = 64$. When a number is multiplied in this way it is said to be *cubed*.

$\sqrt{}$ — Gives the number which achieves your total by being multiplied by itself, i.e. $\sqrt{36} = 6$. This is said to be the *square root* of a number and is the opposite of *squared*.

$\sqrt[3]{}$ — Gives the number which when multiplied by itself three times will be the total. $\sqrt[3]{64} = 4$ this is said to be the *cube root*.

x^{-1} Divides 1 by a number, i.e. $\frac{1}{4} = 0.25$. This is the *reciprocal* button and is useful in this book for finding the resistance of resistors in parallel and capacitors in series.

EXP   The powers of 10 function, i.e.

$$25 \times 1000 = 25\,\text{EXP} \times 10^3 = 25\,000$$

Enter into calculator $25\,\text{EXP}\,3 = 25\,000$. (Do not enter the x or the number 10.)

If a calculation shows 10^{-3}, i.e. 25×10^{-3} enter $25\,\text{EXP} -3 = (0.025)$ (when using EXP if a minus is required use the button $(-)$).

Brackets   These should be used to carry out a calculation within a calculation.

Example calculation:

$$\frac{32}{(0.8 \times 0.65 \times 0.94)} = 65.46$$

Enter into calculator $32 \div (0.8 \times 0.65 \times 0.94) =$

Remember: *Practice makes perfect!*

Simple transposition of formulae

To find an unknown value:

- The subject must be on the top line and must be on its own.

- The answer will always be on the top line.

- To get the subject on its own, values must be moved.

- Any value that moves across the = sign must move

from above the line to below line or from below the line to above the line.

EXAMPLE 1

$$\frac{3 \times 4 = 2 \times 6}{3 \times 4 = 2 \times \,?}$$

Transpose to find ?

$$\frac{3 \times 4}{2} = 6.$$

EXAMPLE 2

$$\frac{2 \times 6}{?} = 4$$

Step 1 $\quad \dfrac{2 \times 6 = 4 \times \,?}{}$

Step 2 $\dfrac{2 \times 6}{4} = ?$

Answer $\dfrac{2 \times 6}{4} = 3.$

EXAMPLE 3

$$5 \times 8 \times 6 = 3 \times 20 \times ?$$

Step 1: Move 3×20 away from the unknown value, as the known values move across the $=$ sign they must move to the bottom of the equation

$$\dfrac{5 \times 8 \times 6}{3 \times 20} = ?$$

Step 2: Carry out the calculation

$$\dfrac{5 \times 8 \times 6}{3 \times 20} = \dfrac{240}{60} = 4$$

Therefore

$$5 \times 8 \times 6 = 240$$

$$3 \times 20 \times 4 = 240$$

or

$$5 \times 8 \times 6 = 3 \times 20 \times 4.$$

SI units

In Europe and the UK, the units for measuring different properties are known as SI units.
SI stands for *Système Internationale*.

All units are derived from seven base units.

Base quantity	Base unit	Symbol
Time	Second	s
Electrical current	Ampere	A
Length	Metre	m
Mass	Kilogram	kg
Temperature	Kelvin	K
Luminous intensity	Candela	cd
Amount of substance	Mole	mol

SI-DERIVED UNITS

Derived quantity	Name	Symbol
Frequency	Hertz	Hz
Force	Newton	N
Energy, work, quantity of heat	Joule	J
Electric charge, quantity of electricity	Coulomb	C
Power	Watt	W
Potential difference, electromotive force	Volt	V or U

Capacitance	Farad	F
Electrical resistance	Ohm	Ω
Magnetic flux	Weber	Wb
Magnetic flux density	Tesla	T
Inductance	Henry	H
Luminous flux	Lumen	cd
Area	Square metre	m^2
Volume	Cubic metre	m^3
Velocity, speed	Metre per second	m/s
Mass density	Kilogram per cubic metre	kg/m^3
Luminance	Candela per square metre	cd/m^2

SI UNIT PREFIXES

Name	Multiplier	Prefix	Power of 10
Tera	1 000 000 000 000	T	1×10^{12}
Giga	1 000 000 000	G	1×10^{9}
Mega	1 000 000	M	1×10^{6}
Kilo	1000	k	1×10^{3}
Unit	1		
Milli	0.001	m	1×10^{-3}
Micro	0.000 001	μ	1×10^{-6}
Nano	0.000 000 001	η	1×10^{-9}
Pico	0.000 000 000 001	ρ	1×10^{-12}

EXAMPLES

mA	Milliamp = one thousandth of an ampere
km	Kilometre = one thousand metres
μv	Microvolt = one millionth of a volt
GW	Gigawatt = one thousand million watts
kW	Kilowatt = one thousand watts

Calculator example

1 kilometre is 1 metre $\times 10^3$

Enter into calculator 1 EXP 3 $= (1000)$ metres

1000 metres is 1 kilometre $\times 10^{-3}$

Enter into calculator 1000 EXP $-3 = (1)$ kilometre

1 microvolt is 1 volt $\times 10^{-6}$

Enter into calculator 1 EXP $-6 = (1^{-06}$ or 0.000001) volts (note sixth decimal place).

Conductor colour identification

	Old colour	*New colour*	*Marking*
Phase 1 of a.c.	Red	Brown	L1
Phase 2 of a.c.	Yellow	Black	L2
Phase 3 of a.c.	Blue	Grey	L3
Neutral of a.c.	Black	Blue	N

Note Great care must be taken when working on installations containing old and new colours.

EXERCISE I

1. Convert 2.768 kW to watts.
2. How many ohms are there in 0.45 MΩ?
3. Express a current of 0.037 A in milliamperes.
4. Convert 3.3 kV to volts.
5. Change 0.000 596 MΩ to ohms.
6. Find the number of kilowatts in 49 378 W.
7. The current in a circuit is 16.5 mA. Change this to amperes.
8. Sections of the 'Grid' system operate at 132 000 V. How many kilovolts is this?
9. Convert 1.68 μC to coulombs.
10. Change 724 mW to watts.
11. Convert the following resistance values to ohms:
    - (a) 3.6 μΩ
    - (b) 0.0016 MΩ
    - (c) 0.085 MΩ
    - (d) 20.6 μΩ
    - (e) 0.68 μΩ

12. Change the following quantities of power to watts:
    (a) 1.85 kW   (d) 1850 µW
    (b) 18.5 mW   (e) 0.0185 kW
    (c) 0.185 MW
13. Convert to volts:
    (a) 67.4 mV   (d) 9250 µV
    (b) 11 kV   (e) 6.6 kV
    (c) 0.240 kV
14. Convert the following current values to amperes:
    (a) 345 mA   (d) 0.5 mA
    (b) 85.4 µA   (e) 6.4 mA
    (c) 29 mA
15. Add the following resistances together and give the answer in ohms: 18.4 Ω, 0.000 12 MΩ, 956 000 µΩ
16. The following items of equipment are in use at the same time: four 60 W lamps, two 150 W lamps, a 3 kW immersion heater, and a 1.5 kW radiator. Add them to find total load and give the answer in watts.
17. Express the following values in more convenient units:
    (a) 0.0053 A   (d) 0.000 006 25 C
    (b) 18 952 W   (e) 264 000 V
    (c) 19 500 000 Ω
18. The following loads are in use at the same time: a 1.2 kW radiator, a 15 W lamp, a 750 W iron, and a 3.5 kW washing machine. Add them together and give the answer in kilowatts.
19. Add 34 250 Ω to 0.56 MΩ and express the answer in ohms.
20. From 25.6 mA take 4300 µA and give the answer in amperes.
21. Convert 32.5 µC to coulombs.
22. Convert 4350 pF to microfarads.
23. 45 µs is equivalent to:
    (a) 0.45 s   (c) 0.0045 s
    (b) 0.045 s   (d) 0.000 045 s
24. 50 cl is equivalent to:
    (a) 51   (b) 0.051   (c) 0.05 ml   (d) 500 ml

25. $0.2 \, \text{m}^3$ is equivalent to:

   (a) $200 \, \text{dm}^3$ **(c)** $2000 \, \text{dm}^3$

   (b) $2000 \, \text{cm}^3$ **(d)** $200 \, \text{cm}^3$

26. $0.6 \, \text{M}\Omega$ is equivalent to:

   (a) $6000 \, \Omega$ **(c)** $600\,000 \, \Omega$

   (b) $60\,000 \, \Omega$ **(d)** $6\,000\,000 \, \Omega$

Areas, perimeters and volumes

AREAS AND PERIMETERS

Rectangle

Fig. 1

To calculate perimeter, add length of all sides, i.e. $3 + 2 + 3 + 2 = 10$ m

To calculate area, multiply the length by breadth, i.e. $3 \times 2 = 6$ m^2

Triangle

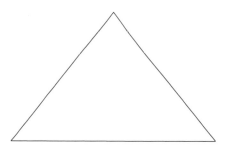

Fig. 2

Area = half base multiplied by height, $1.5 \times 1.6 = 2.4$ m^2

Circle

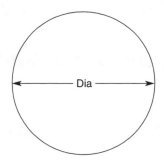

Fig. 3

Circumference $= \pi \times d$ $3.142 \times 80 = 251.36$ mm

If required in m$^2 = \dfrac{251.36}{1000} = 0.251$ m

Area:
$$A = \frac{\pi \times d^2}{4} = \frac{3.142 \times 80 \times 80}{4} = 5027.2 \text{ mm}^2$$

Calculator method:

Enter: shift $\pi \times 80\,x^2 = 5027.2$ mm^2

VOLUME

Diameter 58 mm, height 246 mm

Volume = area of base of cylinder $\times$ height

Base has a diameter of 58 mm.

Area of base $= \dfrac{\pi d^2}{4} = 2642$ mm^2

Volume = area $\times$ height $2642 \times 246 = 649\,932$ mm^3

To convert mm^3 to m^3

Enter into calculator $649\,932$ EXP $-9 = 6.499\,32 \times 10^{-04}$
(thousand times smaller)

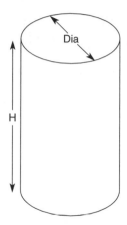

Fig. 4

EXAMPLE 1   To calculate the cross-sectional area of a trunking with dimensions of 50 mm by 100 mm.

$$\text{Area} = \text{length} \times \text{breadth}, \quad 50 \times 100 = 5000 \text{ mm}^2$$

EXAMPLE 2   To calculate the area of a triangular space 6.5 metres wide and 8.6 metres high.

$$\text{Area} = \frac{1}{2} \, b \times h$$

or

$$\frac{1}{2} \times 6.5 \times 8.6 = 27.95 \text{ m}^2$$

Enter on calculator $0.5 \times 6.5 \times 8.6 = 27.95 \text{ m}^2$

EXAMPLE 3   A cylinder has a diameter of 0.6 m and a height of 1.3 m. Calculate its volume and the length of weld around its base.

$$\text{Volume} = \frac{\pi \times d^2}{4} \times \text{height}$$

$$\frac{\pi \times 0.6^2}{4} \times 1.3 = 0.368 \text{ m}^3 \text{ (round up)}$$

Enter on calculator shift $\pi \times 0.6 \, x^2 \div 4 \times 1.3 = (0.367 \, \text{m}^2)$

EXAMPLE 4 Calculate the volume of a rectangular tank with a base 1.2 m long, 600 mm wide, 2.1 m high.

$$1.2 \times (600\,\text{mm convert to metres})\,0.6 \times 2.1 = 1.51\,\text{m}^3$$

Calculate the length of insulation required to wrap around the tank.

$$1.2 + 0.6 + 1.2 + 0.6 = 3.6\,\text{metres}$$

EXERCISE 2

1. Find the volume of air in a room 5 m by 3.5 m by 2.6 m.
2. Calculate the volume of a cylindrical tank 0.5 m in diameter and 0.75 m long.
3. Find the volume and total surface area of the following enclosed tanks:
   (a) rectangular, 1 m × 0.75 m × 0.5 m.
   (b) cylindrical, 0.4 m in diameter and 0.5 m high.
4. Find the volume of a copper bar 6 m long and 25 mm by 8 mm in cross-section.
5. Calculate the volume per metre of a length of copper bar with a diameter of 25 mm.
6. The gable end wall of a building is 15 m wide and 5 m high with the triangular area of the roof being 3 m high. The building is 25 m long. Calculate the volume of the building.
7. A triangular roof has a width of 2.8 m and a height of 3 m. Calculate the volume of the roof if the building was 10.6 m long.
8. Calculate the area of material required to make a cylindrical steel tank with a diameter of 1.2 m and a height of 1.8 m. The calculation is to include lid and base.
9. A storage tank has internal dimensions of 526 mm × 630 mm × 1240 mm. Calculate the volume of the tank allowing an additional 15%.
10. A circular tank has an external diameter of 526 mm and an external length of 1360 mm. It is made from 1.5 mm thick metal. Calculate the volume within the tank.

Space factors

Any cables installed into a trunking or duct should not use more than 45% of the available space (cross-sectional area) within the trunking or duct. This is called the space factor. This can be calculated or, alternatively, tables from the *On-Site Guide* can be used.

Calculation first:

EXAMPLE I   Calculate the amount of usable area within a trunking 50 mm by 75 mm. Cross-sectional area of trunking can be found $50 \times 75 = 3750$ mm^2. 45% of this area can be found

$$\frac{3750 \times 45}{100} = 1687.5 \text{ mm}^2.$$

Enter on calculator 3750×45 shift % $= 1687.5$
This is the amount of space that can be used.

When calculating how many cables can be installed in the trunking, it is important to take into account the insulation around the cable as this counts as used space.

Table A
Details of single core thermoplastic (pvc) cables

Nominal conductor size (mm^2)	Number and diameter of wires (no. of strands $\times$ mm^2)	Nominal overall diameter (mm)
1.0	1×1.13	2.9
1.5	1×1.38	3.1
2.5	1×1.78	3.5
2.5 (stranded)	7.067	3.8
4	7×0.85	4.3

Continued

(Cont'd)

Nominal conductor size (mm^2)	Number and diameter of wires (no. of strands $\times$ mm^2)	Nominal overall diameter (mm)
6	7 × 1.04	4.9
10	7 × 1.35	6.2
16	7 × 1.70	7.3
25	7 × 2.14	9.0
35	19 × 1.53	10.3
50	19 × 1.78	12.0

Table B
Dimensions of trunking (mm × mm)

50 × 37.5
50 × 50
75 × 25
75 × 37.5
75 × 50
75 × 75
100 × 25
100 × 37.5
100 × 50
100 × 75
100 × 100

EXAMPLE 2 Calculate the maximum number of 10 mm^2 cables that could be installed in a 50 mm × 75 mm trunking allowing for space factor.

Find area of trunking 50 × 75 = 3750 mm^2

Usable area (45%) 3750 × 45 shift % = 1687.50 (calculator)

or $\dfrac{3750 \times 45}{100} = 1687.50 \, mm^2$

From Table A, the diameter of a 10 mm^2 cable is 6.2 mm.

The cross-sectional area (csa) of one cable is

$$\frac{\pi d^2}{4} = \frac{3.142 \times 6.2^2}{4} = 30.19 \, mm^2$$

To calculate the number of cables that it would be permissible to install in the trunking.

$$\frac{\text{Usable area}}{\text{csa of cable}} = \text{Number of cables}$$

$$\frac{1687.5}{30.19} = 55.89, \quad \text{therefore 55 cables can be installed.}$$

EXAMPLE 3 The following cables are to be installed in a single run of trunking: $12 \times 1\,\text{mm}^2$, $10 \times 1.5\,\text{mm}^2$, $8 \times 2.5\,\text{mm}^2$ stranded, $6 \times 25\,\text{mm}^2$.

Calculate the size of trunking required for this installation.

Step 1:

Calculate the cross-sectional area of cables using values from Table A

csa of $1\,\text{mm}^2$ cable including insulation $\dfrac{\pi \times 2.9^2}{4} = 6.6\,\text{mm}^2$

Twelve cables: $12 \times 6.6 = 79.2\,\text{mm}^2$

csa of $1.5\,\text{mm}^2$ cable including insulation $\dfrac{\pi \times 3.1^2}{4} = 7.54\,\text{mm}^2$

Ten cables: $10 \times 7.54 = 75.4\,\text{mm}^2$

csa of $2.5\,\text{mm}^2$ cable including insulation $\dfrac{\pi \times 3.8^2}{4} = 11.34\,\text{mm}^2$

Eight cables: $8 \times 11.34 = 90.72\,\text{mm}^2$

csa of $25\,\text{mm}^2$ cable including insulation $\dfrac{\pi \times 9^2}{4} = 63.61\,\text{mm}^2$

Six cables: $6 \times 63.61 = 381.66\,\text{mm}^2$

Step 2:

Add all cross-sectional areas of cables together:

$$79.2 + 75.4 + 90.72 + 381.66 = 629.98\,\text{mm}^2$$

This is the total area required for the cables and it must be a maximum of 45% of total area in the trunking.

Step 3:

Calculate space required $\dfrac{629.98 \times 100}{45} = 1399.9\,\text{mm}^2$

Calculator method $629.98 \times 100 \div 45 = 1399.9\,\text{mm}^2$

A $37.5\,\text{mm} \times 50\,\text{mm}$ trunking has an area of $37.5 \times 50 = 1875\,\text{mm}^2$

This will be suitable and will also allow some space for future additions.

The methods shown above are perfectly suitable for the calculation of space factor and it is necessary to learn these calculations.

However, it is far easier to use tables from the *On-Site Guide* which with practice simplify choosing the correct size trunking.

Appendix 5 of the *On-Site Guide* is for conduit and trunking capacities. For trunking, Tables 5E and 5F should be used.

EXAMPLE 4   A trunking is required to contain the following thermoplastic cables (singles)

$26 \times 1.5 \text{ mm}^2$ stranded

$12 \times 2.5 \text{ mm}^2$ stranded

$12 \times 6 \text{ mm}^2$

$3 \times 10 \text{ mm}^2$

$3 \times 25 \text{ mm}^2$

Calculate the minimum size trunking permissible for the installation of these cables.

From Table 5E, each cable has a factor as follows. Once found, the factors should be multiplied by the number of cables.

$1.5 \text{ mm}^2 = 8.6 \times 26 = 223.6$

$2.5 \text{ mm}^2 = 12.6 \times 12 = 151.20$

$6.0 \text{ mm}^2 = 21.2 \times 12 = 254.4$

$10.0 \text{ mm}^2 = 35.3 \times 3 = 105.9$

$25.0 \text{ mm}^2 = 73.9 \times 3 = 221.7$

Add the cable factors together $= 956.8$

From Table 5F (factors for trunking), a factor larger than 956 must now be found.

It will be seen from the table that a trunking 100×25 has a factor of 993 therefore this will be suitable, although possibly a better choice would be 50×50 which has a factor of 1037 as this will allow for future additions.

It should be remembered that there are no space factors for conduit, the amount of cables that can be installed in a conduit is dependent on the length of conduit and the number of bends between drawing-in points.

Appendix 5 of the *On-Site Guide* contains tables for the selection of single-core insulated cables installed in conduit.

EXAMPLE 5   A conduit is required to contain ten single-core 1.0 mm^2 pvc-insulated cables. The length of conduit between the control switches and an electric indicator-lamp box is 5 m, and the conduit run has two right-angle bends. Select a suitable size of conduit.

From Table 5C (*IEE On-Site Guide*), the cable factor for ten 1.0 mm^2 cables $= 10 \times 16 = 160$.

From Table 5D (*IEE On-Site Guide*), for a 5 m run with two bends select 20 mm conduit with a conduit factor of 196.

EXAMPLE 6   Steel conduit is required to contain the following stranded-conductor single-core pvc-insulated cables for a machine circuit:

(a) three 6 mm^2 cables between a steel cable trunking and the machine control box,

(b) in addition to the 6 mm^2 cables between the control box and the machine there are three 2.5 mm^2 cables and six 1.5 mm^2 cables.

The conduit from the cable trunking to the machine control box is 1.5 m long with one bend, and the steel conduit between the control box and the machine is 3.5 m with two bends.

Using appropriate cable and conduit tables, select suitable conduit sizes for:

(i) trunking to control box and

(ii) control box to machine.

(i) From Table 5A (*IEE On-Site Guide*) cable factor for 6 mm^2 cable is 58, thus cable factor for three 6 mm^2 cables $= 58 \times 3 = 174$.

From Table 5D (*IEE On-Site Guide*) for a conduit run of 2.5 m with one bend select 20 mm conduit with a factor of 278.

(ii) From Table 5C (*IEE On-Site Guide*) cable factor for 6 mm^2 cable is 58, thus cable factor for three 6 mm^2 cables $= 58 \times 3 = 174$. Again from table 5C cable factor for 2.5 mm^2 cable is 30, thus cable factor for three 2.5 mm^2 cables $= 30 \times 3 = 90$ and from table 5C cable factor for 1.5 mm^2 cable is 22, thus cable factor for six 1.5 mm^2 cable $= 22 \times 6 = 132$.

Thus total cable factors $= 174 + 90 + 132 = 396$.

From Table 5D (*IEE On-Site Guide*) for a 3.5 m conduit run with two bends, select 25 mm conduit with a factor of 404.

EXERCISE 3

1. The floor of a room is in the form of a rectangle 3 m by 3.5 m. Calculate its area.
2. A rectangular electrode for a liquid resistor is to have area 0.07 mm^2. If it is 0.5 m long, how wide must it be?
3. Complete the table below, which refers to various rectangles:

Length (m)	6		12	8	
Breadth (m)	2	2			
Perimeter (m)		10		24	32
Area (m^2)			84		48

4. The triangular portion of the gable end of a building is 6 m wide and 3.5 m high. Calculate its area.
5. The end wall of a building is in the form of a square with a triangle on top. The building is 4 m wide and 5.5 high to the top of the triangle. Calculate the total area of the end wall.
6. Complete the table below, which refers to various triangles:

Base (m)	0.5	4	1.5		0.3
Height (m)	0.25		2.2	3.2	0.12
Area (m^2)		9		18	

7. Complete the following table:

Area (m^2)	0.015			0.000 29	0.0016
Area (mm^2)		250	7500		

8. Complete the table below, which refers to various circles:

Diameter	0.5 m		4 mm
Circumference		1.0 m	
Area		0.5 m^2	6 mm^2

9. A fume extract duct is to be fabricated on site from aluminum sheet. Its dimensions are to be 175 mm diameter and 575 mm length. An allowance of 25 mm should be left for a riveted joint along its length. Establish the area of metal required and the approximate number of rivets required, assuming rivets at approximately 70 mm spacing.

10. A square ventilation duct is to be fabricated on site from steel sheet. To avoid difficulty in bending the corners are to be formed by 37.5 mm × 37.5 mm steel angle and 'pop' riveting. Its dimensions are to be 259 mm × 220 mm × 660 mm length. Establish the area of sheet steel, length of steel angle and the approximate number of rivets required, assuming rivets at 60 mm spacing.

11. A coil of wire contains 25 turns and is 0.25 m in diameter. Calculate the length of wire in the coil.

12. Complete the table below, which refers to circular conductors:

Number and diameter of wires (mm)	1/1.13	7/0.85		
Nominal cross-sectional area of conductor (mm^2)		2.5	10	25

13. Complete the table below, which refers to circular cables:

Nominal overall diameter of cable(mm)	2.9	3.8	6.2	7.3	12.0
Nominal overall cross-sectional area (mm^2)					

14. Calculate the cross-sectional areas of the bores of the following heavy-gauge steel conduits, assuming that the wall thickness is 1.5 mm:

    (a) 16 mm   (b) 25 mm   (c) 32 mm

15. Complete the following table, using a space factor of 45% in each case:

	Permitted number of pvc cables in trunking of size (mm)		
Cable size	50 × 37.5	75 × 50	75 × 75
16 mm²			
25 mm²			
50 mm²			

16. The following pvc cables are to be installed in a single run of trunking: twelve 16 mm², six 35 mm², twenty-four 2.5 mm², and eight 1.5 mm².

Determine the size of trunking required, assuming a space factor of 45%.

17. Determine the size of square steel trunking required to contain the following pvc cables: fifteen 50 mm², nine 25 mm², eighteen 10 mm². Take the space factor for ducts as 35%.

18. The nominal diameter of a cable is 6.2 mm. Its cross-sectional area is

(a) 120.8 mm²   **(c)** 30.2 mm²

(b) 19.5 mm²   **(d)** 61.2 mm²

19. Allowing a space factor of 45%, the number of 50 mm² cables that may be installed in a 50 mm × 37.5 mm trunking is

(a) 71   **(b)** 8   **(c)** 23   **(d)** 37

The following cable calculations require the use of data contained in documents based upon BS 7671, e.g. *IEE On-Site Guide*, etc. In each case assume that the stated circuit design calculations and environmental considerations have been carried out to determine the necessary cable current ratings and type of wiring system.

20. A steel cable trunking is to be installed to carry eighteen 1.5 mm² single-core pvc-insulated cables to feed nine floodlighting luminaires; a single 4 mm² protective conductor is to be included in the trunking. Establish the minimum size of trunking required.

21. 50 mm × 38 mm pvc trunking is installed along a factory wall to contain low-current control cables. At present there are 25 pairs of single-core 1.5 mm^2 pvc-insulated cables installed. How many additional pairs of similar 1.5 mm^2 control cables may be installed in the trunking?

22. A pvc conduit is to be installed to contain six 4 mm^2 single-core pvc cables and one 2.5 mm^2 stranded single-core pvc protective conductor. The total length of run will be 16 m and it is anticipated that four right-angle bends will be required in the conduit run. Determine the minimum conduit size and state any special consideration.

23. An electric furnace requires the following wiring:
    (i)   three 6 mm^2 stranded single-core pvc cables,
    (ii) four 2.5 mm^2 stranded single-core pvc cables,
    (iii) four 1.5 mm^2 stranded single-core pvc cables.
    There is a choice between new steel conduit and using existing 50 mm × 38 mm steel trunking which already contains six 25 mm^2 single-core pvc cables and four 10 mm^2 single-core pvc cables. Two right-angle bends will exist in the 18 m run.
    (a) Determine the minimum size of conduit to be used, and
    (b) state whether the new cables could be included within the existing trunking, and if they could be, what considerations must be given before their inclusion.

24. Select two alternative sizes of steel trunking which may be used to accommodate the following.
    (i)   ten 16 mm^2 single-core pvc-insulated cables,
    (ii) twelve 6 mm^2 single-core pvc-insulated cables,
    (iii) sixteen 1.5 mm^2 single-core pvc-insulated cables,
    (iv) three multicore pvc-insulated signal cables, assuming a cable factor of 130.
    An extension to the trunking contains ten of the 16 mm^2 cables and 8 of the 1.5 mm^2 cables.
      Establish the minimum size of conduit, assuming a 5 m run with no bends. How may the conduit size selected affect the choice of trunking dimensions (assume that the two sizes of trunking cost the same).

Coulombs and current flow

Current is a flow of electrons.

When 6 240 000 000 000 000 000 electrons flow in one second a current of one ampere is said to flow. This quantity of electrons is called a coulomb (C) and is the unit used to measure electrical charge.

1 coulomb = 6.24×10^{18} electrons

Therefore 1 coulomb = 1 ampere per second.

The quantity of electrical charge $Q = I \times t$ coulombs.

EXAMPLE 1   Calculate the current flow if 7.1 coulombs were transferred in 2.5 seconds.

$$I = \frac{Q}{t} = \frac{7.1}{2.5} = 2.84\,\text{A}$$

EXAMPLE 2   If a current of 12 A flows for 4.5 minutes, calculate the quantity of electricity that is transferred

$$Q = I \times t$$

$$Q = 12 \times (4.5 \times 60)\,3240\,\text{coulombs}$$

EXERCISE 4

1. Calculate the time taken for a current of 14 A to flow at a charge of 45 C.
2. How long must a current of 0.5 A flow to transfer 60 coulombs?
3. If a current of 4.3 A flows for 15 min, calculate the charge transferred.

Circuit calculations

OHM'S LAW

The symbol used for voltage unit and quantity in the calculations will be U (V can be used if preferred).

U — Voltage can be thought of as the pressure in the circuit

I — Current is the flow of electrons

R (Ω) — Resistance is anything which resists the flow of current, i.e. cable resistance, load resistance or a specific value of resistance added to a circuit for any reason.

In a d.c. circuit, the current is directly proportional to the applied voltage and inversely proportional to the resistance.

The formulae for Ohm's law calculations are:

$$U = I \times R$$

$$R = \frac{U}{I}$$

$$I = \frac{U}{R}$$

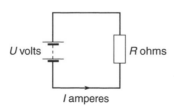

Fig. 5

If a voltage of 100 V is applied to a 5 Ω resistor

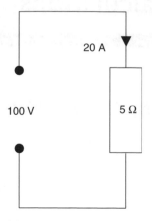

Fig. 6

$$\frac{U}{R} = \frac{100}{5} = 20 \text{ A}$$

If the resistance in the circuit is increased to 10 Ω, it can be seen that the current flow reduces

$$\frac{U}{R} = \frac{100}{10} = 10 \text{ A}$$

double the resistance and the current is halved.
If it was the resistance of the circuit that was unknown, the calculation

$$\frac{U}{I} = R$$

could be used

$$\frac{100}{10} = 10 \text{ Ω}$$

If the voltage was an unknown value, $I \times R = U$ could be used:

$$10 \text{ Ω} \times 10 \text{ A} = 100 \text{ volts}$$

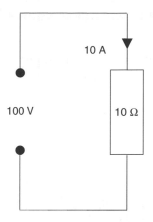

Fig. 7

RESISTORS IN SERIES

When a number of resistors are connected in series, the total resistance is equal to the sum of all of the resistance values.

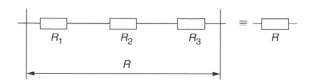

Fig. 8

$$R_1 + R_2 + R_3 = R$$

$$0.413 + 0.23 + 0.6 = 1.243$$

Fig. 9

Ohm's law used to calculate values in series circuits

Using the circuit in Figure 8 with an applied voltage of 10 volts, the total current can be calculated:

$$\frac{U}{R} = I$$

$$\frac{10}{1.243} = 8.05 \text{ A}$$

It can be seen by calculation that each resistance in the circuit will cause a reduction in the voltage (pressure). Ohm's law can be used to find the voltage at different parts of the circuit. (*The current is the common value in a series circuit as it will be the same wherever it is measured.*)

The calculation $I \times R$ can be used to calculate the voltage drop across each resistance.

Using values from Figure 8, the current in the circuit is 8.04 Ω.

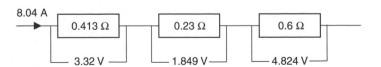

Fig. 10

Voltage drop across R_1 will be $8.04 \times 0.413 = 3.32$ volts

                     R_2 $8.04 \times 0.23 \ = 1.849$ volts

                     R_3 $8.04 \times 0.6 \ \ = 4.824$ volts

                     Total $= 9.993$

It can be seen that the sum of the voltage drop across all resistors is equal to the total voltage in the circuit and that the voltage after the last resistance is 0 volts.

1. Calculate the total resistance of each of the following groups of resistors in series. (*Values are in ohms unless otherwise stated.*)

   (a) 12, 35, 59

   (b) 8.4, 3.5, 0.6

   (c) 19.65, 4.35

   (d) 0.085, 1.12, 0.76

   (e) 27.94, 18.7, 108.3

   (f) 256.5, 89.7

   (g) 1400, 57.9 k Ω

   (h) 1.5 M Ω, 790 000

   (i) 0.0047, 0.095

   (j) 0.0568, 0.000 625 (*give answers in microhms*)

2. Determine the value of resistance which, when connected in series with the resistance given, will produce the required total.

   (a) 92 Ω to produce 114 Ω

   (b) 12.65 Ω to produce 15 Ω

   (c) 1.5 Ω to produce 3.25 Ω

   (d) 4.89 Ω to produce 7.6 Ω

   (e) 0.9 Ω to produce 2.56 Ω

   (f) 7.58 Ω to produce 21 Ω

   (g) 3.47 Ω to produce 10 Ω

   (h) 195 Ω to produce 2000 Ω

   (i) 365 $\mu\Omega$ to produce 0.5 Ω (*answer in microhms*)

   (j) 189 000 Ω 0.25 M Ω (*answer in megohms*)

3. Calculate the total resistance when four resistors each of 0.84 Ω are wired in series.

4. Resistors of 19.5 Ω and 23.7 Ω are connected in series. Calculate the value of a third resistor which will give a total of 64.3 Ω.

5. How many 0.58 Ω resistors must be connected in series to make a total resistance of 5.22 Ω?

6. A certain type of lamp has a resistance of 41 Ω. What is the resistance of 13 such lamps in series? How many of these lamps are necessary to make a total resistance of 779 Ω?

7. The four field coils of a motor are connected in series and each has a resistance of $33.4\,\Omega$. Calculate the total resistance. Determine also the value of an additional series resistance which will give a total resistance of $164\,\Omega$.

8. Two resistors connected in series have a combined resistance of $4.65\,\Omega$. The resistance of one of them is $1.89\,\Omega$. What is the resistance of the other?

9. Four equal resistors are connected in series and their combined resistance is $18.8\,\Omega$. The value of each resistor is
   (a) $9.4\,\Omega$ (b) $75.2\,\Omega$ (c) $4.7\,\Omega$ (d) $37.6\,\Omega$

10. Two resistors connected in series have a combined resistance of $159\,\Omega$. One resistor has a value of $84\,\Omega$. The value of the other is
    (a) $133.56\,\Omega$ (c) $243\,\Omega$
    (b) $1.89\,\Omega$ (d) $75\,\Omega$

11. Two resistors of equal value are connected to three other resistors of value $33\,\Omega$, $47\,\Omega$ and $52\,\Omega$ to form a series group of resistors with a combined resistance of $160\,\Omega$.
    What is the resistance of the two unknown resistors?
    (a) $7\,\Omega$ (b) $14\,\Omega$ (c) $28\,\Omega$ (d) $42\,\Omega$

12. Four resistors of value $23\,\Omega$, $27\,\Omega$, $33\,\Omega$, $44\,\Omega$ are connected in series. It is required to modify their combined resistance to $140\,\Omega$ by replacing one of the existing resistors by a new resistor of value $40\,\Omega$. Which of the original resistors should be replaced?
    (a) $23\,\Omega$ (b) $27\,\Omega$ (c) $33\,\Omega$ (d) $44\,\Omega$

RESISTORS IN PARALLEL

When resistances are connected in parallel, the voltage is common to each resistance. (*Remember in series it was the current that was common.*)

Each resistance which is connected to a circuit in parallel will reduce the resistance of the circuit and will therefore increase the current flowing in the circuit.

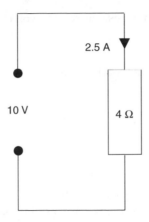

Fig. 11

Figure 11 shows a resistance of 4 Ω connected to a voltage of 10 V. Using ohm's law the current in the circuit can be calculated:

$$\frac{U}{R} = I$$

$$\frac{10}{4} = 2.5 \text{ A}$$

When another resistance of 4 Ω is connected to the circuit in parallel, as Figure 12, the total resistance can be calculated, again by using Ohm's law as follows.

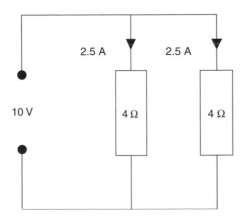

Fig. 12

The voltage across each resistance is 10 volts. Therefore,

The current flowing through R_1 is $\dfrac{10}{4} = 2.5$ A

The current flowing through R_2 is $\dfrac{10}{2} = 5$ A

The total current in the circuit is the sum of the current flowing in R_1 and R_2, i.e. $2.5 + 5 = 7.5$ amperes.

If another resistance of 6 Ω is connected in parallel to this circuit as Figure 13.

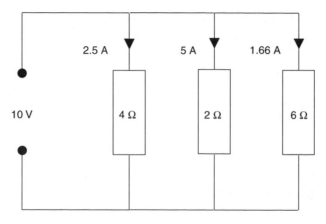

Fig. 13

Current flowing in R_1 is 2.5 A. R_2 is 5 A Using Ohm's law, current in R_3 is

$$\frac{10}{6} = 1.66 \text{ amperes}$$

(**Note** higher resistance results in less current flow.)

Total current in circuit is $2.5 + 5 + 1.66 = 9.16$ amperes

This can also be used to calculate the total resistance of the circuit.

Using Ohm's law, the voltage is 10 volts, current is 9.16, the calculation is

$$\frac{U}{I} = R$$

$$\frac{10}{9.1} = 1.09 \ \Omega$$

Clearly, this method can only be used if the voltage is known.

Calculation of total resistance of resistors in parallel

If only the resistances are known, the total resistance can be calculated by using the following method with a calculator:

$$\frac{1}{R} + \frac{1}{R} + \frac{1}{R} = \frac{1}{R} = R$$

Using values from Figure 13:

$$\frac{1}{4} + \frac{1}{2} + \frac{1}{6} = \frac{1}{1.09} = \text{total resistance}$$

On calculator enter

$$4x^{-1} + 2x^{-1} + 6x^{-1} = x^{-1} =$$

Note, x^{-1} is a function (button) on the calculator.
This can be proven to be correct by using Ohm's law again

$$\frac{U}{R} = I$$

The current will be the same as when the currents passing through all of the individual resistances in Figure 13 are added together, i.e.

$$2.5 + 5 + 1.66 = 9.16 \text{ amperes}$$

Resistances in parallel using product over sum method

Another method of calculating the total resistance of resistances in parallel is by using the product over sum method:

If the resistances from Figure 13 are used, the total resistance can be found:

$$\frac{4 \times 2}{4 + 2} = \frac{8}{6} = \frac{1.333 \times 6}{1.333 + 6} = \frac{7.998}{7.333} = 1.09\,\Omega$$

This method can be used for any number of resistances connected in parallel.

The calculation must be carried out using two resistances, then using the result of the calculation with the next resistance, then the next, until all of the resistances are used.

EXERCISE 6

1. The following groups of resistors are connected in parallel. In each case calculate the equivalent resistance. Where necessary, make the answers correct to three significant figures. (*All values are in ohms.*)

   (a) 2, 3, 6   (f) 14, 70
   (b) 3, 10, 5   (g) 12, 12
   (c) 9, 7   (h) 15, 15, 15
   (d) 4, 6, 9   (i) 40, 40, 40, 40
   (e) 7, 5, 10

2. In each case, calculate the value of a resistor which, when connected in parallel with the given resistor, will produce the value asked for.

   (*Give answers correct to three significant figures.*)

	Given resistance Ω	Resistance required Ω
(a)	48	12
(b)	20	5
(c)	9	4
(d)	6	3

(e)	7	6
(f)	500	400
(g)	0.6×10^3	200
(h)	75	25
(i)	38	19
(j)	52	13

3. A heating element is in two sections, each of $54\,\Omega$ resistance. Calculate the current taken from a $230\,V$ supply when the sections are connected (a) in series, (b) in parallel.

4. Two single-core cables, having resistances of $1.2\,\Omega$ and $0.16\,\Omega$, are connected in parallel and are used to carry a total current of $30\,A$. Calculate (a) the voltage drop along the cables, (b) the actual current carried by each cable.

5. A cable carries a current of $65\,A$ with a $13\,V$ drop. What must be the resistance of a cable which, when connected in parallel with the first cable, will reduce the voltage drop to $5\,V$?

6. To vary the speed of a d.c. series motor it is usual to connect a diverter resistor in parallel with the field winding.

   The field of a series motor has a resistance of $0.6\,\Omega$ and the diverter resistor has three steps, of $5\,\Omega$, $4\,\Omega$ and $2\,\Omega$. Assuming that the total current is fixed at $28\,A$, find out how much current flows through the field winding at each step of the diverter.

7. Resistors of $24\,\Omega$ and $30\,\Omega$ are connected in parallel. What would be the value of a third resistor to reduce the combined resistance to $6\,\Omega$?

8. Two cables having resistances of $0.03\,\Omega$ and $0.04\,\Omega$ between them carry a total current of $70\,A$. How much does each carry?

9. When two equal resistors are connected in series to a $125\,V$ supply, a current of $5\,A$ flows. Calculate the total current which would flow from the same voltage supply if the resistors were connected in parallel.

10. A current of $50\,A$ is carried by two cables in parallel. One cable has a resistance of $0.15\,\Omega$ and carries $20\,A$. What is the resistance of the other cable?

11. Three cables, having resistances of $0.018\,\Omega$, $0.024\,\Omega$ and $0.09\,\Omega$ respectively, are connected in parallel to carry a total current of $130\,A$. Calculate

   (a) the equivalent resistance of the three in parallel,

   (b) the voltage drop along the cables,

   (c) the actual current carried by each cable.

12. Four resistance coils – A, B, C and D – of values $4\,\Omega$, $5\,\Omega$, $6\,\Omega$ and $7\,\Omega$ respectively, are joined to form a closed circuit in the form of a square. A direct-current supply at $40\,V$ is connected across the ends of coil C. Calculate

   (a) the current flowing in each resistor,

   (b) the total current from the supply,

   (c) the potential difference across each coil,

   (d) the total current from the supply if a further resistance coil R of $8\,\Omega$ is connected in parallel with coil A.

13. Resistors of $3\,\Omega$, $5\,\Omega$ and $8\,\Omega$ are connected in parallel. Their combined resistance is
   (a) $1.6\,\Omega$ (b) $0.658\,\Omega$ (c) $16.0\,\Omega$ (d) $1.52\,\Omega$

14. Two resistors are connected in parallel to give a combined resistance of $3.5\,\Omega$. The value of one resistor is $6\,\Omega$. The value of the other is
   (a) $8.4\,\Omega$ (b) $0.12\,\Omega$ (c) $1.2\,\Omega$ (d) $2.5\,\Omega$

15. The resistance of a cable carrying $43\,A$ is $0.17\,\Omega$. Calculate the resistance of a second cable which, if connected in parallel, would reduce the voltage drop to $5\,V$.

16. A cable of resistance $1.92\,\Omega$ carries a current of $12.5\,A$. Find the voltage drop. If a second cable of $2.04\,\Omega$ resistance is connected in parallel, what voltage drop will occur for the same value of load current?

17. Three cables, having resistances $0.0685\,\Omega$, $0.0217\,\Omega$ and $0.1213\,\Omega$, are connected in parallel. Find (a) the resistance of the combination, (b) the total current which could be carried by the cables for a voltage drop of $5.8\,V$.

18. A load current of 250 A is carried by two cables in parallel. If their resistances are 0.0354 Ω and 0.046 Ω, how much current flows in each cable?

19. Two cables in parallel between them carry a current of 87.4 A. One of them has a resistance of 0.089 Ω and carries 53 A. What is the resistance of the other?

20. Resistors of 34.7 Ω and 43.9 Ω are connected in parallel. Determine the value of a third resistor which will reduce the combined resistance to 19 Ω.

21. Three pvc-insulated cables are connected in parallel, and their resistances are 0.012 Ω, 0.015 Ω and 0.008 Ω, respectively, With a total current of 500 A flowing on a 240 V supply,

    (a) calculate the current in each cable,
    (b) calculate the combined voltage drop over the three cables in parallel,
    (c) calculate the individual voltage drop over each cable in the paralleled circuit.

22. Tests on a 300 m length of single-core mineral insulated cable produced the following results: conductor resistance 2.4 Ω, insulation resistance 40 MΩ. What will be the anticipated conductor and insulation resistance values of a 120 m length of the cable?

    (a) 16 Ω, 0.96 MΩ   (c) 0.96 Ω, 40 MΩ
    (b) 0.96 Ω, 16 MΩ   (d) 16 Ω, 16 MΩ

23. A 250 m reel of twin mineral insulated cables is to be cut to provide two equal lengths. Before cutting the cable one core is tested and the insulation resistance is found to be 23 MΩ and the conductor resistance found to be 2.9 Ω. What will be the anticipated conductor and insulation resistance values of each of the two lengths?

    (a) 46 Ω, 1.45 MΩ   (c) 0.145 Ω, 11.5 MΩ
    (b) 1.45 Ω, 46 MΩ   (d) 11.5 Ω, 46 MΩ

EXAMPLE Resistors of $4\,\Omega$ and $5\,\Omega$ are connected in parallel and a $6\,\Omega$ resistor is connected in series with the group. The combination is connected to a 100 volt supply (Figure 14). Calculate the total resistance, voltage drop and current in each resistor.

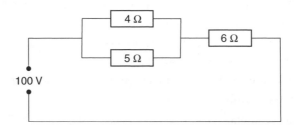

Fig. 14

To find a resistance for the parallel group

$$\frac{1}{R_1} + \frac{1}{R_2} = \frac{1}{R} = R$$

$$\frac{1}{4} + \frac{1}{5} = \frac{1}{0.45} = 2.22\,\Omega \quad \text{(calculator method)}$$

Circuit may now be shown as in Figure 15.

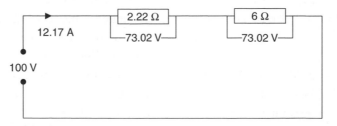

Fig. 15

Total resistance in circuit can now be calculated as two resistances in series.

Total $R = 2.22 + 6 = 8.22\,\Omega$

To calculate total current
Using Ohm's law

$$\frac{U}{R} = I$$

$$\frac{100}{8.22} = 12.17\,\text{A}$$

Voltage drop across the 6 Ω resistance is calculated

$$I \times R = U$$

$$12.17 \times 6 = 73.02\,\text{V}$$

Voltage drop across parallel group is 100 V − 73.02 V = 26.98 V

This voltage can now be used to calculate the current through each parallel resistance, again using Ohm's law.

Current through 4 Ω resistor is

$$\frac{V}{R} = I$$

$$\frac{26.98}{4} = 6.745\,\text{A}$$

Current through 5 Ω resistor is

$$\frac{26.98}{5} = 5.396\,\text{A}$$

As a check the sum of the currents through the parallel resistances together should equal the total current in the circuit

6.745 + 5.396 = 12.141 (allowing for only using three decimal places).

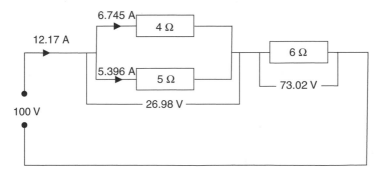

Fig. 16

Internal resistance

EXAMPLE   A heating element of $2.4\,\Omega$ resistance is connected to a battery of e.m.f. $12\,\text{V}$ and internal resistance $0.1\,\Omega$ (Figure 17). Calculate

(a)   the current flowing,
(b)   the terminal voltage of the battery on load,
(c)   the power dissipated by the heater.

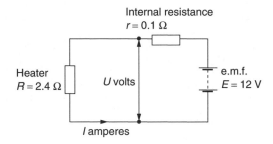

Internal resistance
$r = 0.1\,\Omega$

Heater
$R = 2.4\,\Omega$

U volts

e.m.f.
$E = 12\,\text{V}$

I amperes

Fig. 17

(a)   Using E for e.m.f. and U for terminal voltage, and treating the internal resistance as an additional series resistance,

$$\text{e.m.f.} = \text{total current} \times \text{total resistance}$$

$$E = I \times (R + r) \quad \text{(note the use of brackets)}$$

$$\therefore \qquad 12 = I \times (2.4 + 0.1)$$

$$= I \times 2.5$$

$$\therefore \qquad I = \frac{12}{2.5}$$

$$= 4.8\,\text{A}$$

(b) The terminal voltage is the e.m.f. minus the voltage drop across the internal resistance:

$$\text{terminal voltage } U = E - Ir$$
$$= 12 - (4.8 \times 0.1)$$
$$= 12 - 0.48$$
$$= 11.52 \quad \text{or} \quad 11.5 \text{ V}$$

(c) The power dissipated in the heater is

$$P = U \times I$$
$$= 11.5 \times 4.8$$
$$= 55 \text{ W}$$

EXERCISE 7

1. For the circuit of Figure 18, find
   - **(a)** the resistance of the parallel group,
   - **(b)** the total resistance,
   - **(c)** the current in each resistor.

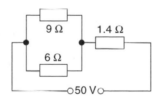

Fig. 18

2. For the circuit of Figure 19, find
   - **(a)** the total resistance,
   - **(b)** the supply voltage.

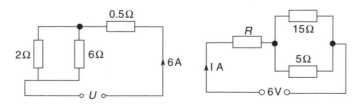

Fig. 19   **Fig. 20**

3. Find the value of the resistor R in the circuit of Figure 20.
4. Calculate the value of the resistor r in the circuit of Figure 21.

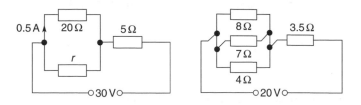

Fig. 21   **Fig. 22**

5. For the circuit of Figure 22, find
   (a) the total resistance,
   (b) the total current.
6. Determine the voltage drop across the 4.5 Ω resistor in Figure 23.

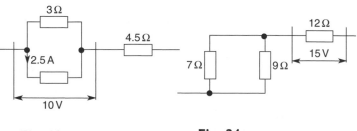

Fig. 23   **Fig. 24**

7. Calculate the current in each resistor in Figure 24.
8. Determine the value of a resistor which when connected in parallel with the 70 Ω resistor will cause a total current of 2.4 A to flow in the circuit of Figure 25.

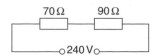

Fig. 25

9. Two contactor coils of resistance 350 Ω and 420 Ω, respectively, are connected in parallel. A ballast resistor of 500 Ω is connected in series with the pair. Supply is taken from a 220 V d.c. supply. Calculate the current in each coil and the power wasted in the ballast resistor.

10. Two 110 V lamps are connected in parallel. Their ratings are 150 W and 200 W. Determine the value of a resistor which when wired in series with the lamps will enable them to operate from the 230 V mains.

11. A shunt motor has two field coils connected in parallel, each having a resistance of 235 Ω. A regulating resistor is wired in series with the coils to a 200 V supply. Calculate the value of this resistor when the current through each coil is 0.7 A.

12. In a certain installation the following items of equipment are operating at the same time: (i) a 3 kW immersion heater, (ii) two 100 W lamps, (iii) one 2 kW radiator. All these are rated at 240 V.

    The nominal supply voltage is stated to be 230 V but it is found that the actual voltage at the origin of the installation is 5 V less than this. Calculate
    (a) the total current,
    (b) the resistance of the supply cables,
    (c) the actual power absorbed by the immersion heater.

13. The overhead cable supplying an outbuilding from the 230 V mains supply has a resistance of 0.9 Ω. A 2 kW radiator and a 1500 W kettle, both rated at 230 V, are in use at the time. Determine the voltage at the terminals of this apparatus. What would be the voltage if a 750 W, 240 V water heater were also switched on?

14. Two resistors in parallel, A of 20 Ω and B of unknown value, are connected in series with a third resistor C of 12 Ω. The supply to the circuit is direct current.

    If the potential difference across the ends of C is 180 V and the power in the complete circuit is 3600 W, calculate
    (a) the value of resistor B,
    (b) the current in each resistor,
    (c) the circuit voltage.

15. State Ohm's law in your own words, and express it in symbols. A d.c. supply at 240 V is applied to a circuit comprising two resistors A and B in parallel, of 5 Ω and 7.5 Ω, respectively, in series with a third resistor C of 30 Ω.

    Calculate the value of a fourth resistor D to be connected in parallel with C so that the total power in the circuit shall be 7.2 kW.

16. Three resistors of value 1.5 Ω, 4 Ω and 12 Ω, respectively, are connected in parallel. A fourth resistor, of 6 Ω, is connected in series with the parallel group. A d.c. supply of 140 V is applied to the circuit.

    (a) Calculate the current taken from the supply.

    (b) Find the value of a further resistor to be connected in parallel with the 6 Ω resistor so that the potential difference across it shall be 84 V.

    (c) What current will now flow in the circuit?

17. An electric bell takes a current of 0.3 A from a battery whose e.m.f. is 3 V and internal resistance 0.12 Ω. Calculate the terminal voltage of the battery when the bell is ringing.

18. Determine the voltage at the terminals of a battery of three cells in series, each cell having an e.m.f. of 1.5 V and internal resistance 0.11 Ω, when it supplies a current of 0.75 A.

19. A car battery consists of six cells connected in series. Each cell has an e.m.f. of 2 V and internal resistance of 0.008 Ω. Calculate the terminal voltage of the battery when a current of 105 A flows.

20. A battery has an open-circuit voltage of 6 V. Determine its internal resistance if a load current of 54 A reduces its terminal voltage to 4.35 V.

21. Resistors of 5 Ω and 7 Ω are connected in parallel to the terminals of a battery of e.m.f. 6 V and internal resistance of 0.3 Ω. Calculate

    (a) the current in each resistor.

    (b) the terminal voltage of the battery,

    (c) the power wasted in internal resistance.

22. A battery is connected to two resistors, of 20 Ω and 30 Ω, which are wired in parallel. The battery consists of three cells in series, each cell having an e.m.f. of 1.5 V and internal resistance 0.12 Ω. Calculate

    (a) the terminal voltage of the battery,

    (b) the power in each resistor.

23. A battery of 50 cells is installed for a temporary lighting supply. The e.m.f. of each cell is 2 V and the internal resistance is 0.0082 Ω. Determine the terminal voltage of the battery when it supplies 25 lamps each rated at 150 W, 110 V.

24. The installation in a country house is supplied from batteries. The batteries have an open-circuit voltage of 110 V and an internal resistance of 0.045 Ω. The main cables from the batteries to the house have a resistance of 0.024 Ω. At a certain instant the load consists of two 2 kW radiators, three 100 W lamps, and four 60 W lamps. All this equipment is rated at 110 V. Calculate the voltage at the apparatus terminals.

25. An installation is supplied from a battery through two cables in parallel. One cable has a resistance of 0.34 Ω, the other has a resistance of 0.17 Ω. The battery has an internal resistance of 0.052 Ω and its open-circuit voltage is 120 V. Determine the terminal voltage of the battery and the power wasted in each cable when a current of 60 A is flowing.

26. A 12 V battery needs charging and the only supply available is one of 24 V. The battery has six cells, each of e.m.f. 1.8 V and internal resistance 0.009 Ω. Determine the value of a series resistor which will limit the current to 5 A.

27. A circuit consists of a 7.2 Ω resistor in parallel with one of unknown value. This combination is connected in series with a 4.5 Ω resistor to a supply of direct current. The current flowing is 2.2 A and the total power taken by the circuit is 35 W. Calculate

    (a) the value of the unknown resistor,

    (b) the supply voltage.

(c) the value of a resistor which if connected in parallel
with the 4.5 Ω resistor will cause a current of 4 A
to flow.

(Assume that the source of supply has negligible internal
resistance.)

28. The combined resistance of the circuit in Figure 26 is
    (a) 0.333 Ω **(b)** 12.5 Ω **(c)** 30.0 Ω **(d)** 7.7 Ω

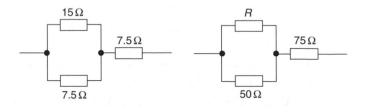

Fig. 26   **Fig. 27**

29. The combined resistance of the circuit in Figure 27 is
    91.7 Ω. The value of resistor R is
    (a) 33.3 Ω **(b)** 250 Ω **(c)** 0.04 Ω **(d)** 25 Ω
30. The current flowing in the 0.4 Ω resistor in Figure 28 is
    (a) 8.57 A **(b)** 11.43 A **(c)** 0.24 A **(d)** 0.73 A

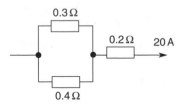

Fig. 28

Resistivity

The resistance of a conductor is

$$R = \frac{\rho \times l}{A} \,\Omega$$

ρ is the resistivity (Ωm)
l is the conductor length (m)
A is the cross-sectional area (m^2)

EXAMPLE 1   Determine the resistance of 100 m of 120 mm^2 copper.

The resistivity of copper is 1.78×10^{-8} Ω

$$R = \frac{\rho \times l}{A}$$

$$\frac{1.78 \times 10^{-8} \times 100}{120 \times 10^{-6}} = 0.0148 \,\Omega$$

(note 10^{-6} to convert to sq m)

Enter into calculator 1.78 EXP $-8 \times 100 \div 120$ EXP $-6 = 0.0148$

EXAMPLE 2   Calculate the length of a 2.5 mm^2 copper conductor that will have a resistance of 1.12 Ω.

This requires the use of simple transposition, it is easier to start with the formulae that is known:

$$R = \frac{\rho \times l}{A}$$

Replace the letters with numbers where possible

$$1.12 = \frac{1.78 \times 10^{-8} \times L}{2.5 \times 10^{-6}}$$

The subject must be on its own on the top line, this will require moving some of the values.

Remember when a value moves across the = sign it must move from bottom to top or top to bottom. This will give us:

1st step

$$2.5 \times 10^{-6} \times 1.12 = 1.78 \times 10^{-8} \times L$$

2nd step

$$\frac{2.5 \times 10^{-6} \times 1.12}{1.78 \times 10^{-8}} = L$$

This will leave L on its own and we can now carry out the calculation Enter into calculator

$$2.5\,\text{EXP} - 6 \times 1.12 \div 1.78\,\text{EXP} - 8 = 157.30\,\text{m}$$

EXAMPLE 3   Calculate the cross-sectional area of an aluminium cable 118 m long which has a resistance of 0.209 Ω.

The resistivity of aluminium is 2.84×10^{-8}

$$R = \frac{\rho \times l}{A}$$

Convert to values

$$0.209 = \frac{2.84 \times 10^{-8} \times 118}{A \times 10^{-6}}$$

Transpose

$$A \times 0.209 = \frac{2.84 \times 10^{-8} \times 118}{10^{-6}}$$

$$A = \frac{2.84 \times 10^{-8} \times 118}{0.209 \times 10^{-6}} = 16\,\text{mm}^2$$

Enter into calculator

$$2.84\,\text{EXP} - 8 \times 118 \div 0.209 \times \text{EXP} - 6 = 16.03\,\text{mm}^2$$

In the following exercises, take the resistivity of copper as $1.78 \times 10^{-8} \, \Omega m$ and that of aluminium as $2.84 \times 10^{-8} \, \Omega m$.

1. Determine the resistance of 100 m of copper cable whose cross-sectional area is $1.5 \, mm^2$.

2. Calculate the resistance of 50 m of copper cable $4 \, mm^2$ in cross-sectional area.

3. Find the cross-sectional area of a copper cable which is 90 m long and has a resistance of $0.267 \, \Omega$.

4. Find the cross-sectional area of a copper cable 42 m long which carries a current of 36 A with a voltage drop of 2.69 V.

5. A certain grade of resistance wire has a resistivity of $50 \times 10^{-8} \, \Omega m$. Find the length of this wire needed to make a heating element of resistance $54 \, \Omega$. Assume the cross-sectional area of the wire is $0.4 \, mm^2$.

6. Calculate the voltage drop produced in a 75 m length of twin copper cable $16 \, mm^2$ in cross-sectional area when it carries 25 A. What would be the voltage drop if the same size of aluminium cable were used?

7. Determine the resistance of 30 m of copper busbar 60 mm by 6 mm.

8. Calculate the thickness of an aluminium busbar 60 mm wide and 12 m long which has a resistance of $0.000946 \, \Omega$.

9. Find the resistance of 35 m of $1 \, mm^2$ copper cable.

10. Calculate the voltage drop produced by a current of 40 A in 24 m of single $10 \, mm^2$ copper cable.

11. Find the length of resistance wire 1.2 mm in diameter needed to construct a $20 \, \Omega$ resistor. (Resistivity $= 50 \times 10^{-8} \, \Omega m$.)

12. Find the resistance of 125 m of $50 \, mm^2$ aluminium cable.

13. A $6 \, mm^2$ copper twin cable carries a current of 32 A, and there is a voltage drop of 4.5 V. Calculate the length of the cable.

14. Iron is sometimes used for making heavy-duty resistors. Its resistivity is $12 \times 10^{-8} \, \Omega m$. Calculate the resistance of an iron grid, the effective length of which is 3 m and which is 10 mm by 6 mm in cross-section.

15. Calculate the diameter of an aluminium busbar which is 24 m long and whose resistance is $0.00139\,\Omega$.

16. A d.c. load current of 28 A is to be supplied from a point 30 m away. Determine a suitable cross-sectional area for the copper cable in order that the voltage drop may be limited to 6 V.

17. Calculate the resistance per 100 m of the following sizes of copper cable:

    (a) $1.5\,\text{mm}^2$ (d) $35\,\text{mm}^2$

    (b) $6\,\text{mm}^2$ (e) $50\,\text{mm}^2$

    (c) $10\,\text{mm}^2$

18. The resistance of 1000 m of a certain size of cable is given as $0.168\,\Omega$. Find by the method of proportion the resistance of (a) 250 m, (b) 180 m, (c) 550 m.

19. The resistance of a certain length of a cable of cross-sectional area $2.5\,\text{mm}^2$ is $5.28\,\Omega$. Find by the method of proportion the resistance of a similar length of cable whose cross-sectional area is (a) $10\,\text{mm}^2$, (b) $25\,\text{mm}^2$, (c) $1.5\,\text{mm}^2$.

20. The following figures refer to a certain size of cable:

Length (m)	1000	750	500	250
Resistance (Ω)	0.4	0.3	0.2	0.1

    Plot a graph to show the relationship between resistance and length (length horizontally, resistance vertically) and read from the graph the resistance of a 380 m length of the cable.

21. The following table shows the resistance of cables having the same length but different cross-sectional areas:

c.s.a. (mm^2)	1.0	1.5	2.5	4	10
Resistance (Ω)	15.8	12.4	8	3.43	2.29

    Plot a graph to show the relationship between resistance and cross-sectional area (cross-sectional area horizontally, resistance vertically). From the graph, find the resistance of a cable whose cross-sectional area is $6\,\text{mm}^2$.

22. The resistance of 100 m of $2.5\,\text{mm}^2$ copper cable is

    (a) $0.712\,\Omega$ (c) $1.404\,\Omega$

    (b) $7.12\,\Omega$ (d) $0.0712\,\Omega$

23. The resistance of 15 m of aluminium bar 60 mm by 7.5 mm in cross-section is

    (a) $0.0946\,\Omega$ **(c)** $9.46 \times 10^{-4}\,\Omega$

    (b) $9.46 \times 10^{-3}\,\Omega$ **(d)** $1.58 \times 10^{-2}\,\Omega$

24. A shunt for an ammeter is required to have a resistance of $0.002\,\Omega$. If made of copper strip 100 mm long, the cross-sectional area of the strip would be

    (a) $0.89\,\text{mm}^2$ **(c)** $1.12\,\text{mm}^2$

    (b) $8.9\,\text{mm}^2$ **(d)** $11.2\,\text{mm}^2$

Voltage drop

CONDUCTOR RESISTANCE AND VOLTAGE DROP USING OHM'S LAW

Regulation number 525-01-02 states that the maximum voltage drop in any circuit from the origin of the supply, to the terminals of the current using equipment must not exceed 4% of the supply voltage.

If supply voltage is 230 V, the calculation to find 4% is

$$\frac{230 \times 4}{100} = 9.2 \text{ V}$$

Enter into calculator $230 \times 4\% = 9.2$

Therefore, total voltage drop permissible is 9.2 volts on a 230 volt supply.

As described in the chapter on series resistances, there will be a voltage drop across any resistances in series. A conductor will be a resistance in series with the resistance of a load.

This voltage drop can be calculated using Ohm's law.

EXAMPLE A circuit is wired using $70°C$ thermoplastic flat twin and earth cable with copper 2.5 mm^2 live conductors and a 1.5 mm^2 circuit protective conductor, the circuit is 30 metres long and will carry a current of 17 amperes, supply voltage is 230 volts.

From Table 9A in the *On-Site-Guide*, it can be seen that a 2.5 mm^2 copper conductor has a resistance of 7.41 mΩ per metre at $20°C$.

The current flowing in a circuit will be the same in the phase and the neutral conductors (see Figure 29). Therefore the resistance of both live conductors must be taken into account.

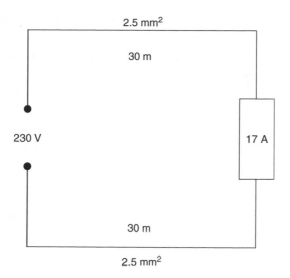

Fig. 29

Phase conductor resistance is 7.41 mΩ/M
Neutral conductor resistance is 7.41 mΩ/M

$$7.41 + 7.41 = 14.82$$

The resistance of a twin 2.5 mm² copper cable is 14.82 mΩ/M, this can also be found using Table 9A of *On-Site Guide*.
The total resistance of this cable will be mΩ per metre × length.

$$14.82 \times 30 = 444.6\,m\Omega$$

This value is in milli-ohms and should now be converted to ohms:

$$\frac{m\Omega}{1000} = ohms$$

$$\frac{432.6}{1000} = 0.444\,\Omega$$

When conductors are operating their maximum current rating they can operate at 70°C. This will result in the resistance of the conductors increasing. This increased resistance must be used in the calculation for voltage drop.

To calculate the total resistance of the cables at their operating temperature a factor from Table 9c in the *On-Site Guide* should be used. It will be seen that a multiplier of 1.2 should be used for a conductor rated at 70°C.

To calculate the total resistance of the current carrying conductors:

$$\Omega \times \text{multiplier} = \text{total resistance of conductors at } 70°\text{C}$$

$$0.444 \times 1.2 = 0.533 \ \Omega$$

These calculations can be carried out in one single calculation:

$$\frac{\text{M}\Omega \times \text{length} \times \text{multiplier}}{1000} = \text{Total resistance}$$

$$\frac{14.82 \times 30 \times 1.2}{1000} = 0.533$$

Voltage drop can now be calculated using Ohm's law:

$$I \times R = U$$

$$17 \times 0.533 = 9.06 \text{ volts}$$

This voltage drop will be acceptable as it is below 9.2 volts.

Voltage drop using BS 7671

Using the same example, a circuit is wired using 70°C thermoplastic flat twin and earth cable with copper 2.5 mm² live conductors and a 1.5 mm² circuit protective conductor. The circuit is 30 metres long and will carry a current of 17 amperes, supply voltage is 230 volts.

The voltage drop for this cable can be found using Table 4D5A from appendix 4 of BS 7671 or Table 6F in the *On-Site Guide*.

Using either of these tables it will be seen that the voltage drop for 2.5 m² copper cable is 18 mv/A/m millivolts × amperes × distance in metres. (*As value is in millivolts, it must be divided by 1000 to convert to volts.*)

Voltage drop for example circuit is

$$\frac{18 \times 17 \times 30}{1000} = 9.18$$

It can be seen that the voltage drop is slightly higher than when Ohm's law was used in the previous calculation. This is because the voltage drop value used in BS 7671 has been rounded up for ease of calculation.

Power in d.c. and purely resistive a.c. circuits

Power (watts) = voltage (volts) × current (amperes)

$$P = U \times I$$

EXAMPLE 1 The current in a circuit is 4.8 A when the voltage is 240 V. Calculate the power.

$$P = U \times I$$

$$= 240 \times 4.8$$

$$= 1152 \, W$$

EXAMPLE 2 Calculate the current flowing when a 2 kW heater is connected to a 230 V supply.

$$P = U \times I$$

$$2000 = 230 \times I$$

$$\therefore \quad I = \frac{2000}{230}$$

$$= 8.7 \, A$$

EXAMPLE 3   The current in a certain resistor is 15 A and the power absorbed is 200 W. Find the voltage drop across the resistor.

$$P = U \times I$$

$$200 = U \times 15$$

$$\therefore \quad U = \frac{200}{15}$$

$$= 13.3 \, \text{V}$$

EXERCISE 9

1. Complete the following table:

P (watts)			3000	1600	1000		1000	2350	
I (amperes)		6				150	0.2		4.5
U (volts)	240		250	240		100	220	460	240

2. The voltage drop in a cable carrying 12.5 A is 2.4 V. Calculate the power wasted.

3. A d.c. motor takes 9.5 A from a 460 V supply. Calculate the power input to the motor.

4. Calculate the current that flows when each of the following pieces of equipment is connected to the 230 V mains:

   (a) 3 kW immersion heater   **(f)** 60 W lamp

   (b) 1500 W kettle   **(g)** 100 W lamp

   (c) 450 W electric iron   **(h)** 2 kW radiator

   (d) 3.5 kW washing machine   **(i)** 750 W water heater

   (e) 7 kW cooker   **(j)** 15 W lamp

5. Calculate the voltage drop in a resistor passing a current of 93 A and absorbing 10 kW.

6. A cable carries a current of 35 A with a 5.8 V drop. Calculate the power wasted in the cable.

7. A heater is rated at 4.5 kW, 240 V. Calculate the current it takes from

   (a) a 240 V supply   **(b)** a 220 V supply.

8. A motor-starting resistor passes a current of 6.5 A and causes a voltage drop of 115 V. Determine the power wasted in the resistor.

9. Determine the current rating of the resistance wire which would be suitable for winding the element of a 1.5 kW, 250 V heater.

10. Calculate the current taken by four 750 W lamps connected in parallel to a 230 V main.

11. A faulty cable joint causes an 11.5 V drop when a current of 55 A is flowing. Calculate the power wasted at the joint.

12. Two lamps, each with a rating of 100 W at 240 V, are connected in series to a 230 V supply. Calculate the current taken and the power absorbed by each lamp.

13. Determine the current rating of the cable required to supply a 4 kW immersion heater from a 230 V mains.

14. A generator delivers a current of 28.5 A through cables having a total resistance of 0.103 Ω. The voltage at the generator terminals is 225 V. Calculate

   (a) the power generated,

   (b) the power wasted in the cables,

   (c) the voltage at the load.

15. Calculate the value of resistance which when connected in series with a 0.3 W, 2.5 V lamp will enable it to work from a 6 V supply.

16. A motor takes a current of 15.5 A at a terminal voltage of 455 V. It is supplied through cables of total resistance 0.32 Ω. Calculate

   (a) the voltage at the supply end,

   (b) the power input to the motor,

   (c) the power wasted in the cables.

17. Two coils, having resistances of 35 Ω and 40 Ω, are connected to a 100 V d.c. supply (a) in series, (b) in parallel. For each case, calculate the power dissipated in each coil.

18. Two cables, having resistances of 0.036 Ω and 0.052 Ω, are connected in parallel to carry a total current of 190 A. Determine the power loss in each cable.

19. If the power loss in a resistor is 750 W and the current flowing is 18.5 A, calculate the voltage drop across the resistor. Determine also the value of an additional series

resistor which will increase the voltage drop to 55 V when the same value of current is flowing. How much power will now be wasted in the original resistor?

20. A d.c. motor takes a current of 36 A from the mains some distance away. The voltage at the supply point is 440 V and the cables have a total resistance of $0.167\,\Omega$. Calculate

    (a) the voltage at the motor terminals,

    (b) the power taken by the motor,

    (c) the power wasted in the cables,

    (d) the voltage at the motor terminals if the current increases to 42 A.

21. The voltage applied to a circuit is 240 V, and the current is 3.8 A. The power is

    (a) 632 W   (c) 912 W

    (b) 63.2 W   (d) 0.016 W

22. The power absorbed by a heating element is 590 W at a p.d. of 235 V. The current is

    (a) 13 865 A   (c) 0.34 A

    (b) 2.51 A   (d) 25.1 A

23. A faulty cable joint carries a current of 12.5 A, and a voltage drop of 7.5 V appears across the joint. The power wasted at the joint is

    (a) 1.67 W   (c) 93.8 W

    (b) 0.6 W   (d) 60 W

24. A heating element absorbs 2.5 kW of power and the current is 10.5 A. The applied voltage is

    (a) 238 V   (c) 2.38 V

    (b) 26.3 V   (d) 4.2 V

METHOD 2

Power = current^2 × resistance

$$P = I^2 R$$

EXAMPLE 1   Calculate the power absorbed in a resistor of $8\,\Omega$ when a current of $6\,\text{A}$ flows.

$$P = I^2 R$$
$$= 6^2 \times 8$$
$$= 36 \times 8$$
$$= 288\,\text{W}$$

EXAMPLE 2   A current of $12\,\text{A}$ passes through a resistor of such value that the power absorbed is $50\,\text{W}$. What is the value of this resistor?

$$P = I^2 R$$
$$50 = 12^2 \times R$$
$$\therefore \qquad R = \frac{50}{12 \times 12}$$
$$= 0.347\,\Omega$$

EXAMPLE 3   Determine the value of current which when flowing in a resistor of $400\,\Omega$ causes a power loss of $1600\,\text{W}$.

$$P = I^2 R$$
$$\therefore \qquad 1600 = I^2 \times 400$$
$$\therefore \qquad I^2 = \frac{1600}{400} = 4$$
$$\therefore \qquad I = \sqrt{4} = 2\,\text{A}$$

EXERCISE 10

1. Complete the following table:

Power (W)		200		1440	1000	2640	100	
Current (A)	10	5	15		4.2		0.42	1.3
Resistance (Ω)	15		8	10		20		25

2. A current of 20 A flows in cable of resistance 0.325 Ω. Calculate the power loss.

3. Determine the power loss in a cable having a resistance of 0.14 Ω when passing a current of 14.5 A.

4. Determine the value of current which, when flowing in a 40 Ω resistor, dissipates 1000 W.

5. An earth fault current of 38 A passes through a conduit joint which has a resistance of 1.2 Ω. Calculate the power dissipated at the joint.

6. A 100 W lamp passes a current of 0.42 A. Calculate its resistance.

7. In a certain installation the *total* length of cable is 90 m and the resistance of this type of cable is 0.6 Ω per 100 m. Determine (a) the voltage drop, (b) the power loss, when a current of 11.5 A flows.

8. A resistor used for starting a d.c. motor has a value of 7.5 Ω. Calculate the power wasted in this resistor when a starting current of 8.4 A flows.

9. Determine the current rating of resistance wire which would be suitable for a 1000 W heater element of resistance 2.5 Ω.

10. An ammeter shunt carries a current of 250 A and its resistance is 0.000 95 Ω. Calculate the power absorbed by the shunt.

11. What is the resistance of an electric-iron element of 450 W rating and which takes a current of 1.9 A?

12. A joint in a cable has a resistance of 0.045 Ω. Calculate the power wasted at this joint when a current of 37.5 A flows.

13. The resistance measured between the brushes of a d.c. motor is 2.3 Ω. Calculate the power loss in the armature when the current is 13.5 A.

14. Determine the rating in watts of a 1100 Ω resistor which will carry 15 mA.

15. Calculate the maximum current which a 250 Ω resistor rated at 160 W will carry.

$$\text{Power} = \frac{\text{voltage}^2}{\text{resistance}}$$

$$P = \frac{U^2}{R}$$

EXAMPLE 1   Calculate the power absorbed by a $40\,\Omega$ resistor when connected to a $240\,\text{V}$ d.c. supply

$$\begin{aligned} \text{Power absorbed}\ \ P &= \frac{U^2}{R} \\ &= \frac{240 \times 240}{40} \\ &= 1440\,\text{W} \end{aligned}$$

EXAMPLE 2   Determine the resistance of a heater which absorbs $3\,\text{kW}$ from a $240\,\text{V}$ d.c. supply.

$$P = \frac{U^2}{R}$$

$$3000 = \frac{240^2}{R}$$

$$\therefore \qquad \frac{1}{3000} = \frac{R}{240^2}$$

$$\therefore \qquad R = \frac{240 \times 240}{3000} = 19.2\,\Omega$$

EXAMPLE 3   Determine the voltage which must be applied to a $9.8\,\Omega$ resistor to produce $500\,\text{W}$ of power.

$$P = \frac{U^2}{R}$$

$$500 = \frac{U^2}{9.8}$$

$$\therefore \qquad U^2 = 9.8 \times 500$$

$$= 4900$$

$$\therefore \qquad U = \sqrt{4900}$$

$$= 70\,\text{V}$$

EXERCISE 11

1. A contactor coil has resistance of $800\,\Omega$. Calculate the power absorbed by this coil from a $240\,\text{V}$ d.c. supply.
2. A piece of equipment creates a voltage drop of $180\,\text{V}$ and the power absorbed by it is $240\,\text{W}$. Determine its resistance.
3. Calculate the resistance of a $36\,\text{W}$, $12\,\text{V}$ car headlamp bulb.
4. Determine the voltage to be applied to a $6\,\Omega$ resistor to produce $2400\,\text{W}$ of power.
5. Complete the following table:

Power (W)		100	60	125		1800			36
Voltage (V)	80	240	250		240	220	3.5		
Resistance (Ω)	50			20	75			0.29	4

6. Calculate the maximum voltage which may be applied to a $45\,\Omega$ resistor rated at $5\,\text{W}$.
7. Determine the power absorbed by the field coils of a $460\,\text{V}$ d.c. motor. The resistance of the coils is $380\,\Omega$.
8. Determine the resistance of a $230\,\text{V}$, $1\,\text{kW}$ heater.
9. The voltage drop in a cable of resistance $0.072\,\Omega$ is $3.5\,\text{V}$. Calculate the power wasted in the cable.
10. Determine the resistance of a $110\,\text{V}$, $75\,\text{W}$ lamp.
11. The following items of equipment are designed for use on a $240\,\text{V}$ supply. Calculate the resistance of each item.
    (a) $2\,\text{kW}$ radiator
    (b) $3\,\text{kW}$ immersion heater
    (c) $3.5\,\text{kW}$ washing machine
    (d) $450\,\text{W}$ toaster
    (e) $60\,\text{W}$ lamp
    (f) $7\,\text{kW}$ cooker
    (g) $100\,\text{W}$ lamp
    (h) $1500\,\text{W}$ kettle
    (i) $750\,\text{W}$ water heater
    (j) $4\,\text{kW}$ immersion heater
12. Calculate the voltage drop in a resistor of $12.5\,\Omega$ when it is absorbing $500\,\text{W}$.
13. The power dissipated in a $57\,\Omega$ resistor is $1000\,\text{W}$. Determine the current.

14. Two lamps are connected in series to a 200 V supply. The lamps are rated at 150 W, 240 V and 60 W, 240 V. Calculate

    (a) the current taken from the supply

    (b) the total power.

15. Two 1000 W, 240 V heater elements are connected to a 240 V d.c. supply (a) in series (b) in parallel. Calculate

    (a) the combined resistance in each case,

    (b) the power absorbed in each case.

16. Cables of resistance 0.35 Ω and 0.082 Ω are connected in parallel and they share a load of 100 A. Determine the current and power loss in each.

17. The element of an immersion heater has a total resistance of 76.8 Ω and is centre-tapped. Calculate the power absorbed from a 240 V supply when the element sections are (a) in series (b) in parallel.

18. Complete the following table and then plot a graph of power (*vertically*) against current (*horizontally*). Try to make the axes of the graph of equal length, and join the points with a smooth curve.

Power (W)		250	360	400	600	
Current (A)	0.8	2.5		3.15		4.9
Resistance (Ω)	40		40		40	40

    From the graph, state

    (a) what power would be dissipated in a 40 Ω resistor by a current of 3.7 A,

    (b) how much current is flowing when the power is 770 W?

19. Complete the following table and plot a graph of power against voltage. Join the points with a smooth curve.

Power (W)		2000		750	420	180
Voltage (V)	240	200	180	120		
Resistance (Ω)	19.2		19.5		19.1	19.2

    (a) Read off the graph the voltage when the power is 1500 W.

    (b) Extend the graph carefully and find from it the power when the voltage is 250 V.

20. The voltage applied to the field circuit of a motor can be varied from 250 V down to 180 V by means of a shunt field regulator. The resistance of the field coils is 360 Ω. Plot a graph showing the relationship between the power and the applied voltage.

21. A cable of resistance 0.07 Ω carries a current which varies between 0 and 90 A. Plot a graph showing the power loss in the cable against the load current.

22. A current of 4.8 A flows in a resistor of 10.5 Ω. The power absorbed is
    (a) 529.2 W   (c) 2420 W
    (b) 24 192 W   (d) 242 W

23. The power developed in a resistor of 24 Ω is 225 W. The current flowing is
    (a) 9.68 A   (b) 3.06 A   (c) 0.327 A   (d) 30.6 A

24. The resistance of a 110 V, 100 W lamp is
    (a) 1210 Ω (b) 0.011 Ω (c) 8.26 Ω (d) 121 Ω

25. The voltage to be applied to a resistor of 55 Ω in order to develop 50 watts of power is
    (a) 0.95 V   (b) 166 V   (c) 52.4 V   (d) 1.05 V

Mechanics

MOMENT OF FORCE

A force F newtons applied at right angles to a rod of length l metres pivoted at P (Figure 30) produces a turning moment M, where

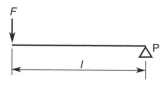

Fig. 30

$$M = F \times l \text{ newton metres (Nm)}$$

(**Note** This turning moment is produced whether or not the bar is actually free to turn.)

EXAMPLE I A horizontal bar 0.5 m long is arranged as in Figure 30. Calculate the force required in order to produce a moment of 250 Nm.

$$M = F \times l$$

$\therefore$ $\qquad 250\,\text{Nm} = F \times 0.5\,\text{m}$

$\therefore$ $\qquad F = \dfrac{250\,\text{Nm}}{0.5\,\text{m}}$

$$= 500\,\text{N}$$

EXAMPLE 2 A horizontal bar 0.75 m long is pivoted at a point
0.5 m from one end, and a downward force of 100 N is applied at
right angles to this end of the bar. Calculate the downward force
which must be applied at right angles to the other end in order to
maintain the bar in a horizontal position. Neglect the weight of
the bar.

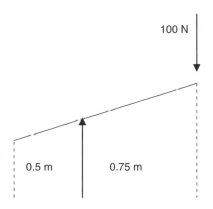

100 N

0.5 m   0.75 m

Fig. 31

The principle of moments applies; that is, for equilibrium (see
Figure 31),

$$\text{total clockwise moment} = \text{total anticlockwise moment}$$

$$F \times 0.25 = 100 \times 0.5$$

$$\therefore \frac{100 \times 0.5}{0.25}$$

$$= 200\,\text{N}$$

(Principle of levers is twice the distance, half the force.)

FORCE RATIO

If the bar of example 2 is considered as a lever, then an *effort* of
100 N is capable of exerting a force of 200 N on an object. The
force F is then in fact the *load*.

By the principle of moments,

$$\text{Load} \times \text{distance from pivot} = \text{effort} \times \left(\begin{array}{c} \text{distance} \\ \text{from pivot} \end{array} \right)$$

The force ratio is $\dfrac{load}{effort}$

Or force ratio $= \dfrac{\text{load}}{\text{effort}} = \dfrac{\text{distance from } effort \text{ to pivot}}{\text{distance from } load \text{ to pivot}}$

In the case of example 2

$$\text{Force ratio} = \frac{0.5\,\text{m}}{0.25} = 2$$

Note that force ratio is often also referred to as 'mechanical advantage'.

MASS, FORCE AND WEIGHT

Very often the load is an object which has to be raised to a higher level against the force of gravity.

The force due to gravity acting on a mass of 1 kg is 9.81 N. The force to raise a mass of 1 kg against the influence of gravity is therefore 9.81 N, and this is called the weight of the 1 kg mass.

Although the newton is the correct scientific unit of force and weight, for industrial and commercial purposes it is usual to regard a mass of 1 kg as having a weight of 1 kilogram force (kgf), therefore

$$1\,\text{kgf} = 9.81\,\text{N}$$

The kilogram force is the 'gravitational' unit of weight and is often abbreviated to 'kilogram', or even 'kilo', in common usage.

EXAMPLE   A crowbar is arranged as shown in Figure 32 and for practical purposes the formula for force ratio may be applied to find the effort required to raise its load of 65 kgf:

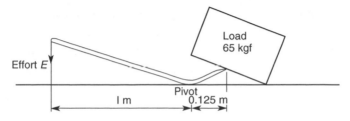

Fig. 32

$$\frac{\text{Load}}{\text{Effort}} = \frac{\text{distance from effort to pivot}}{\text{distance from load to pivot}}$$

$$\frac{65}{E} = \frac{1\,\text{m}}{0.125\,\text{m}}$$

$$\frac{E}{65} = \frac{0.125}{1}$$

$$E = 65 \times 0.125$$

$$= 8.125\,\text{kgf}$$

(or quite simply one eighth of the force)

WORK

When a force F newtons produces displacement of a body by an amount s metres in the direction of the force, the work done is

$$W = F \times s \text{ newton metres or joules (J)}$$

Work = Force × Distance

This is also the energy expended in displacing the body.

EXAMPLE I A force of 200 N is required to move an object through a distance of 3.5 m. Calculate the energy expended.

$$W = F \times s$$

$$= 200\,\text{N} \times 3.5\,\text{m}$$

$$= 700\,\text{Nm or } 700\,\text{J}$$

EXAMPLE 2 Calculate the energy required to raise a mass of 5 kg through a vertical distance of 12.5 m.

We have seen above that the force required to raise a mass of 1 kg against the influence of gravity is 9.81 N; therefore, the force required to raise a mass of 5 kg is

$$F = 5 \times 9.81 \, \text{N}$$

and the energy required is

$$W = 5 \times 9.81 \, \text{N} \times 12.5 \, \text{m}$$

$$= 613 \, \text{Nm or } 613 \, \text{J}$$

THE INCLINED PLANE

Figure 33 illustrates a method of raising a load G through a vertical distance h by forcing it up a sloping plane of length l using an effort E.

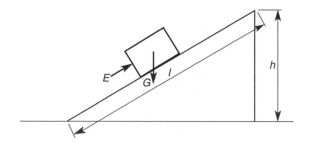

Fig. 33

Ignoring the effects of friction (which can be reduced by using rollers under the load),

$$\text{energy expended by the effort} = \begin{pmatrix} \text{energy absorbed} \\ \text{by the load} \end{pmatrix}$$

that is,

$$\text{effort} \times \begin{pmatrix} \text{distance through} \\ \text{which the effort} \\ \text{is exerted} \end{pmatrix} = \text{load} \times \begin{pmatrix} \text{vertical distance} \\ \text{through which} \\ \text{the load is} \\ \text{raised} \end{pmatrix}$$

$$E \times l = G \times h$$

$$\text{Force ratio} = \frac{\text{load}}{\text{effort}} = \frac{G}{E} = \frac{l}{h}$$

EXAMPLE A motor weighing 100 kgf is to be raised through a vertical distance of 2 m by pushing it up a sloping ramp 5 m long. Ignoring the effects of friction, determine the effort required.

$$\frac{G}{E} = \frac{l}{h}$$

$$\frac{100}{E} = \frac{5}{2}$$

$$\frac{E}{100} = \frac{2}{5}$$

$$E = 100 \times \frac{2}{5}$$

$$= 40\,\text{kgf}$$

THE SCREWJACK

A simplified arrangement of a screw type of lifting jack is shown in cross-section in Figure 34. A horizontal effort E is applied to the arm of radius r and this raises the load G by the action of the screw thread T.

If the effort is taken through a complete revolution, it acts through a distance equal to $2\pi \times r$ (or $\pi \times d^2 \div 4$) and the load rises through a vertical distance equal to the pitch of the screw thread, which is the distance between successive turns of the thread.

If p is the pitch of the thread, and ignoring friction,

$$\begin{pmatrix} \text{energy expended} \\ \text{by the effort} \end{pmatrix} = \begin{pmatrix} \text{energy absorbed by} \\ \text{the load in rising} \\ \text{through a distance } p \end{pmatrix}$$

$$E \times 2\pi r = G \times p$$

$$= \frac{E \times 2\pi r}{p}$$

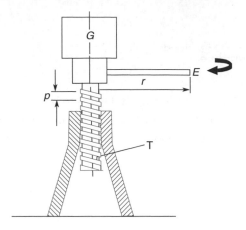

Fig. 34

The force ratio is

$$\frac{\text{load}}{\text{effort}} = \frac{G}{E} = \frac{2\pi r}{p}$$

EXAMPLE   If the pitch of the thread of a screwjack is 1 cm and the length of the radius arm is 0.5 m, find the load which can be raised by applying a force of 20 kg.

$$\frac{G}{E} = \frac{2\pi r}{p}$$

$$\frac{G}{20} = \frac{2\pi\,0.5}{1/100\,\text{m}} \quad \text{(conversion from cm to m)}$$

$$G = \frac{20 \times 2\pi \times 0.5}{0.01}$$

$$= 6283\,\text{kgf}$$

(this gives an enormous advantage but would be very slow.)

THE WHEEL AND AXLE PRINCIPLE

Figure 35 shows a simplified version of a common arrangement by means of which a load G is raised by applying an effort E.

By the principle of moments,

$$E \times R = G \times r$$

$$\text{Force ratio} = \frac{\text{load}}{\text{effort}} = \frac{G}{E} = \frac{R}{r}$$

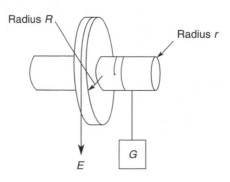

Radius *R*

Radius *r*

E

G

Fig. 35

EXAMPLE   Calculate the effort required to raise a load of 250 kgf using the arrangement shown in Figure 35, if the radius of the large wheel is 20 cm and the radius of the axle is 8 cm.

$$\frac{G}{E} = \frac{R}{r}$$

$$\frac{250}{E} = \frac{20\,\text{cm}}{8\,\text{cm}}$$

$$\frac{E}{250} = \frac{8}{20}$$

$$E = 250 \times \frac{8}{20}$$

$$= 100\,\text{kgf}$$

THE BLOCK AND TACKLE

When a system of forces is in *equilibrium,* the sum of all forces acting downwards is equal to the sum of all forces acting upwards.

Figure 36(a), (b), (c) and (d) illustrates various arrangements of lifting tackle (rope falls) raising a load G by exerting an effort E. In each case the effort is transmitted throughout the lifting rope, giving rise to increasing values of force ratio. (The effects of friction are ignored.)

EXAMPLE Determine the load which (ignoring friction) could be raised by exerting an effort of 50 kgf using each of the arrangements illustrated in Figure 36.

$$\text{For (a), } G = E$$
$$= 50\,\text{kgf}$$
$$\text{For (b), } G = 2E$$
$$= 2 \times 50$$
$$= 100\,\text{kgf}$$
$$\text{For (c), } G = 3E$$
$$= 3 \times 50$$
$$= 150\,\text{kgf}$$
$$\text{For (d), } G = 4E$$
$$= 4 \times 50$$
$$= 200\,\text{kgf}$$

POWER

Power is the rate of doing work

$$\text{Power} = \frac{\text{work done (force} \times \text{distance})}{\text{time taken (time in seconds)}} \text{ or work done in 1 s}$$

EXAMPLE 1 The force required to raise a certain load through a vertical distance of 15 m is 50 N and the operation takes 30 s.

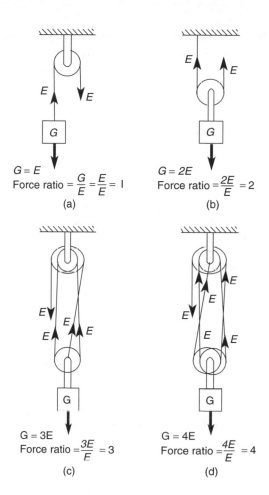

Fig. 36

Calculate the power required

$$\text{Power} = \frac{\text{work done}}{\text{time taken}}$$

$$1\,\text{watt} = 1\,\text{joule of work per second}$$

$$= \frac{50\,\text{N} \times 15\,\text{m}}{35\,\text{s}}$$

$$= \frac{750\,\text{J}}{35\,\text{s}}$$

$$= 21.42\,\text{W}$$

EXAMPLE 2 Calculate the power required to raise a mass of 8 kg through a vertical distance of 23 m in a time of 20 s.

Convert mass to weight 1 kg = 9.81 N (This is the force of gravity on 1 kg.)

$$8 \times 9.81 = 78.48\,\text{N}$$

$$\text{Work done (J)} = \text{force} \times \text{distance}$$

$$78.48 \times 23 = 1805.04$$

$$\text{Power} = \frac{1805.04}{20} = 90.25\,\text{W}$$

or as one calculation

$$\frac{8 \times 9.81 \times 23}{20} = 90.25\,\text{watts}$$

EXAMPLE 3 Calculate the power required to raise 15 m^3 of water per minute through a vertical distance of 35 m. (1 litre of water has a mass of 1 kg.)

The mass of 1 m^3 (1000 litres) of water is 10^3 kg.

The force required to raise this mass of water is

$$F = 0.15 \times 10^3 \times 9.81\,\text{N}$$

$$\text{The power required} = \frac{\text{force} \times \text{distance}}{\text{time in secs}} = \frac{\text{Work}}{\text{time}}$$

as one calculation

$$\frac{0.15 \times 10^3 \times 9.81\,\text{N} \times 35\,\text{m}}{60\,\text{s}} = \frac{\text{Nm}}{\text{s}} \text{ or } \frac{\text{J}}{\text{s}} = 85\,873\,\text{W}$$

$$= 85.9\,\text{kW}$$

Enter into calculator

$$0.15 \text{ EXP } 3 \times 9.81 \times 35 \div 60 =$$

If the pump performing the operation of the last example has an efficiency of 72%. The power required to drive the pump is then

$$P = \frac{85.9\,\text{kW} \times 100}{72} = 119.3\,\text{kW}$$

Enter into calculator

$$85.9 \times 100 \div 72 =$$

EXAMPLE 1   A d.c. motor has a full load output of 5 kW. The input to the motor is 250 V and a current of 26 A is drawn from the supply.

Calculate the efficiency.

$$\text{Efficiency}\,\eta = \frac{\text{output power}}{\text{input power}} \times 100$$

Output power = 5000 W
Input power = volts × amperes 250 × 26 = 6500

$$\text{Efficiency} = \frac{5000 \times 100}{6500} = 76.9\%$$

EXAMPLE 2   Calculate the current taken by a 10 kW 460 V d.c. motor with an efficiency of 78%.

Output power = 10 kW for the calculation this should be converted to watts, i.e. 10 000 W

The input power will always be greater than the output power.

$$\text{Input power}\,\frac{10\,000 \times 100}{78} = 12\,820\,\text{W}$$

To find current drawn from the supply

$$I = \frac{P}{U}$$

$$\frac{12\,820}{460} = 27.86\,\text{A}$$

A simpler method would be

$$I = \frac{p \times 100}{U \times \%}$$

$$\frac{10\,000 \times 100}{460 \times 78} = 27.86\,\text{A}$$

1. A force of 120 N is applied at right angles to the end of a bar 1.75 m long. Calculate the turning moment produced about a point at the other end of the bar.
2. Calculate the force required which when applied at right angles to the end of a bar 0.72 m long will produce a turning moment of 150 Nm about a point at the other end.
3. Complete the following table, which refers to Figure 37:

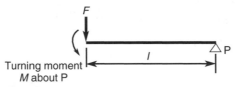

Turning moment
M about P

Fig. 37

F (newtons)	85		0.25	6.5
I (metres)	0.35	1.2		0.125| 2.75
M (Newton metres)		50	0.15		500

4. A bar 1.5 m long is pivoted at its centre. A downward force of 90 N is applied at right angles 0.2 m from one end. Calculate the downward force to be applied at right angles to the bar at the opposite end to prevent it from rotating. Neglect the weight of the bar.
5. A bar 0.8 m long is pivoted at its centre. A downward force of 150 N is applied at right angles to the bar at one end. At what distance from the opposite end of the bar should a vertically downwards force of 200 N be applied to create equilibrium? Neglect the weight of the bar.

6. A force of 25 N is used to move an object through a distance of 1.5 m. Calculate the work done.

7. Energy amounting to 250 J is available to more an object requiring a force of 12.5 N. Through what distance will the object move?

8. Calculate the energy required to raise a load of 240 kg through a vertical distance of 8.5 m.

9. Calculate the energy required to raise 2.5 m³ of water from a well 12.5 m deep.

10. A force of 0.15 N is used to move an object through 75 mm in 4.5 s. Calculate (a) the work done, (b) the power.

11. Calculate the power required to raise a load of 120 kg through a vertical distance of 5.5 m in 45 s.

12. Complete the following table, which refers to Figure 32, page 69

Distance between effort and pivot (m)	1		1.5	1.25			1.8
Distance between load and pivot (m)		0.125			0.15	0.10	0.20
Load (kgf)			200			390	225
Effort (kgf)	20			50	65		
Force ratio			5				

13. Complete the following table, which refers to Figure 33, page 70

Load to be raised (kgf)	250	320	420		500
Effort required (kgf)		150	75	80	
Vertical height (m)	3	4		2.4	1.8
Length of inclined plane (m)	6		5	5.4	4.2

14. A screwjack as illustrated in Figure 34, page 72, has a thread of pitch 8 mm and a radius arm of length 0.5 m. Determine

   (a) the effort required to raise a load of 1000 kgf,

   (b) the load which an effort of 5.5 kgf will raise.

   (c) What length of radius arm would be required to raise a load of 2500 kgf using an effort of 7.5 kgf?

15. Complete the following table, which refers to the wheel and axle illustrated in Figure 35, page 73:

Radius of wheel R (cm)	25	16		17.5	30
Radius of axle r (cm)	8	6.5	6		8.5
Load G (kgf)	200		255	150	175
Effort E (kgf)		75	76.5	72.9	

16. A load of 275 kgf is to be raised using rope falls as illustrated in Figure 36, page 75. Determine the effort required using each of the arrangements (b), (c), and (d). (Ignore friction).

17. An effort of 85 kgf is applied to each of the arrangements in Figure 36(b), (c), and (d), page 75. Ignoring friction, determine the load which could be raised in each case.

18. A motor and gear unit weighing 450 kgf is to be raised through a vertical distance of 2.5 m. It is proposed to use an inclined plane 4 metres long and a set of rope falls as in Figure 36(d). Ignoring friction, determine the effort required.

19. A pump raises $0.15 \, \text{m}^3$ of water per minute from a well 7.5 m deep. Calculate
    (a) the power output of the pump;
    (b) the power required to drive the pump, assuming an efficiency of 75%;
    (c) the energy supplied to the pump in one hour.

20. A test on a d.c. motor produced the following results:
    Input   240 V   Output 3200 W
             15 A
    Calculate the efficiency.

21. Calculate the full-load current of the d.c. motors to which the following particulars refer:

	Supply e.m.f. (V)	Output power (kW)	Efficiency (%)
(a)	240	1	68
(b)	480	15	82
(c)	200	2	74
(d)	250	4	75
(e)	220	10	78

22. A pump which raises $0.12 \, \text{m}^3$ of water per minute through a vertical distance of 8.5 m is driven by a 240 V d.c. motor. Assuming that the efficiency of the pump is 72% and that of the motor is 78%, calculate the current taken by the motor.

23. A motor-generator set used for charging batteries delivers 24 A at 50 V. The motor operates from a 220 V supply and its efficiency is 70%. The generator is 68% efficient. Calculate the cost of running the set per hour at full load if the electrical energy costs 4.79 p per unit.

24. A pumping set delivers $0.6\,m^3$ of water per minute from a well 5 m deep. The pump efficiency is 62%, that of the motor is 74%, and the terminal voltage is 234 V. Calculate

    (a) the motor current;

    (b) the cost of pumping $100\,m^3$ of water with energy at 5.18 p per unit;

    (c) the cross-sectional area of the copper cable which will supply the set from a point 50 m away with a voltage drop of not more than 6 V. (The resistivity of copper is $1.78 \times 10^{-8}\,\Omega m$.)

25. A d.c. motor at 460 V is required to drive a hoist. The load to be raised is 4000 kg at a speed of 0.2 m/s.

    Calculate the minimum power of motor needed to do this work and also the current it would take, assuming the respective efficiencies of hoist gearing and motor to be 85% and 70%.

    State the type of motor to be used, and give reasons for the choice.

26. A 50 m length of two-core cable of cross-section $70\,mm^2$ supplies a 240 V, 30 kW d.c. motor working at full load at 85% efficiency.

    (a) Calculate the voltage drop in the cable.

    (b) What steps would you take to reduce the voltage drop to half the above value, with the same load?

    The resistivity of copper may be taken as $1.78 \times 10^{-8}\,\Omega m$.

27. A conveyor moves 400 kg upwards through a vertical distance of 14 m in 50 s. The efficiency of the gear is 38%. Calculate the power output of the driving motor. The motor is 78% efficient. Calculate the current it takes from a 250 V d.c. supply.

28. The bar in Figure 38 is in equilibrium. The force F is

    (a) 4.8 N   **(b)** 2083 N   **(c)** 208.3 N   **(d)** 75 N

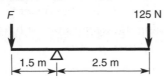

Fig. 38

29. A machine weighing 150 kgf is raised through a vertical distance of 1.5 m by forcing it up a sloping ramp 2.5 m long. Neglecting friction, the effort required is

    (a) 37.5 kgf   **(c)** 250 kgf

    (b) 90 kgf   **(d)** 562.5 kgf

30. With reference to Figure 35, page 73, if the radius of the large wheel is 25 cm and that of the axle is 8.5 cm, the load which could be raised by exerting an effort of 95 kgf is

    (a) 2794 kgf **(b)** 279 kgf   **(c)** 32.3 kgf **(d)** 323 kgf

Power factor

In a purely resistive a.c. circuit, the power drawn from the supply is generally the same as the energy produced at the load.

For example, a 1 kW electric fire will draw 1 kW of power from the supply and produce 1 kW of heat from the fire. This is because the current and voltage are in phase with each other (working together).

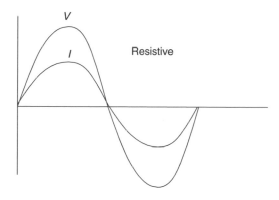

Fig. 39

If we introduce inductance (magnetic effect) into the circuit, as we would in an electric motor, the voltage and current would be out of phase with each other, as the inductance would hold back the current.

This would be known as a 'lagging circuit'.

If we introduced capacitance (electrostatic effect) into the circuit, the voltage and current would be out of phase, it would have the opposite effect to inductance and the voltage would be held back. This would be known as a 'leading circuit'.

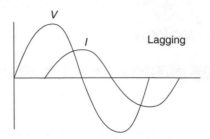

Fig. 40

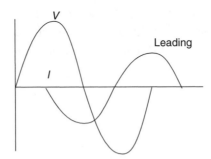

Fig. 41

This happens because inductance or capacitance introduces reactance into the circuit. This is referred to as kVAr (reactive volt amperes).

The effect of reactance on the circuit is that more power is drawn from the supply than is required. This is referred to as kVA (input power).

We already know that output power is referred to in kW (output power).

The power factor is the ratio between the kVA and the kW

$$\frac{\text{kW}}{\text{kVA}} = pf = \frac{\text{Real power (what we get)}}{\text{Apparent power (what we think we should get)}}$$

EXAMPLE A single phase induction motor has an input power of 12.3 kVA and an output power of 11 kW. Calculate the power factor

$$pf = \frac{kW}{kVA}$$

$$\frac{11}{12.3} = 0.89$$

Power factor does not have a unit. It is just a number and will always be less than 1.

A purely resistive circuit has no power factor and is known as unity 1.

Transformers

Transformer calculations can be carried out using the ratio method or by transposition.

The formula is

$$\frac{Up}{Us} = \frac{Np}{Ns} = \frac{Is}{Ip}$$

(connection to supply always made on the primary side)

Up is the voltage on the primary winding; Ip is the current at the secondary winding; Np is the number of primary turns; Us is the voltage at the secondary winding; Is is the current at the secondary winding; Ns is the number of secondary turns.

A step-up transformer is one which has more windings on the secondary side than on the primary side and therefore increases the voltage. A step down transformer is one which has fewer windings on the secondary side than on the primary side and therefore reduces the voltage. The type of transformer can be shown as a ratio. (**Note** We always refer to what happens to the voltage in using the terms step-up and step-down. This is because the current does the opposite.)

EXAMPLE I   A transformer that has 1000 primary turns and 10 000 secondary turns

The ratio is found

$$\frac{Ns}{Np} = \frac{10\,000}{1000} = 10$$

as it has more secondary turns than primary it must be a step-up transformer and the ratio is shown 1:10

If the transformer had 10 000 primary turns and 1000 secondary turns the calculation would be the same. However, because it has fewer secondary turns than it has primary turns, it is a step-down transformer and would be shown as having a ratio of 10:1 (**Note** ratio: primary first:secondary last).

EXAMPLE 2   A transformer has 27 000 turns on the primary winding and 900 turns on the secondary. If a voltage of 230 V was applied to the primary side, calculate
(a) The transformer ratio
(b) The secondary voltage
(a) The ratio is

$$\frac{Up}{Us} \text{ or } \frac{27\,000}{900} = 30$$

as the secondary turns are fewer than the primary it must be a step-down transformer with a ratio of 30:1.
(b)   As the transformer is step-down with a ratio of 30:1, the secondary voltage will be 30 times less than the primary voltage.

$$\frac{230}{30} = 7.66\,\text{V}$$

EXAMPLE 3   A transformer has a step-up ratio of 1:16 it has 32 000 turns on the secondary winding.
   Calculate
(a) The number of turns on the primary winding
(b) The secondary voltage if 50 V is supplied to the primary winding
(a) $\dfrac{Ns}{16} = \dfrac{32\,000}{16} = 2000$ turns

(b)   Using the ratio $50 \times 16 = 800$ volts

TRANSFORMER CURRENT

The ratio of the transformer is the same for current, although when carrying out transformer calculations it must be remembered:

If the voltage is stepped up the current is stepped down.

If the voltage is stepped down the current is stepped up.

$$\frac{Up}{Us} = \frac{Is}{Ip}$$

Using the values from example 3.

If the current supplied from its secondary side is 6 A

$$\frac{50}{800} = \frac{6}{Ip}$$

Using the ratio if it is a step-up voltage transformer, the current will step down by the same ratio.

If the secondary current is 6 A the primary current is:

$$\text{Secondary current} \times \text{ratio} = \text{primary current}$$

$$6 \times 16 = 96\,\text{A}$$

If transposition is used:

Ip must be on its own on the top line

$$\frac{50}{800} = \frac{6}{Ip}$$

Step 1: $\quad \dfrac{50}{} = \dfrac{6 \times 800}{Ip}$

Step 2: $\quad \dfrac{6 \times 800}{50\,Ip}$

Step 3: $\quad \dfrac{Ip}{} = \dfrac{6 \times 800}{50} = 96\,\text{A}$

Electromagnetic effect

MAGNETIC FLUX AND FLUX DENSITY

The unit of magnetic flux is the *weber* (Wb). A magnetic field has a value of 1 Wb if a conductor moving through it in one second has an e.m.f. of 1 volt induced in it.

Convenient units used are milliweber (mWb)

$$1\,\text{Wb} = 10^3\,\text{mWb}$$

and the microweber (µWb)

$$1\,\text{Wb} = 10^6\,\text{µWb}$$

The symbol for magnetic flux is Φ.
The flux density in tesla (symbol B) is calculated by dividing the total flux by the c.s.a. of the magnetic field.

$$B = \frac{\Phi}{A}$$

Φ is the total magnetic flux (Wb); A is the c.s.a. of the magnetic field (m^2); B is the flux density (Wb/m^2 or tesla T).

EXAMPLE I   The total flux in the air gap of an instrument is 0.15 mWb and the c.s.a. of the gap is 500 m^2.

Calculate the flux density (tesla)
Φ in the calculation is in webers, we must convert milliwebers to webers by dividing by 1000 or multiply by (10^{-3})
A is the c.s.a. of the field in m^2. We must convert mm^2 to m^2, as there are one million mm^2 in 1 m^2 we must divide by 1 000 000 or multiply by (10^{-6}).

This can be carried out most simply in one calculation:

$$B = \frac{\Phi}{A}$$

$$\frac{0.15}{1000 \times 500 \times 10^{-6}} = 0.3 \text{ answer in teslas}$$

Enter into calculator (note use of brackets)

$$0.15 \div (1000 \times 500 \, \text{EXP} - 6) =$$

EXAMPLE 2   The air gap in a contactor is 15 mm², the flux density is 1.2 T. Calculate the total flux.
Total area = 15 × 15 = 225 mm²
This requires simple transposition

$$B = \frac{\Phi}{A}$$

or

$$1.2 = \frac{\Phi}{225 \times 10^{-6}}$$

when transposed:

$$225 \times 10^{-6} \times 1.2 = \Phi$$

$(2.7^{-04}$ webers or 0.00027 webers which is 0.27 mW)

FORCE ON A CONDUCTOR WITHIN A MAGNETIC FIELD

When a current carrying conductor is placed at right angles to a magnetic field, the force can be calculated by

$$F = BIl$$

(**Note** It is taken for granted that each letter has a multiplication sign between it and the next letter)
where F is the force in newtons (N), B is the flux density (T), l is the effective conductor length (m) and I is the current (A).

EXAMPLE I   A conductor 300 mm long is placed in and at right angles to a magnetic field with a flux density of 0.5 tesla. Calculate the force exerted on the conductor when a current of 36 A is passed through it.

$$F = B \times I \times l$$

$$F = 0.5 \times 0.3 \times 36 \text{ (note conversion of mm to m)} = 5.4\,\text{N}$$

EXAMPLE 2   A conductor 200 mm long is placed in and at right angles to a magnetic field with a flux density of 0.35 tesla. Calculate the current required in the conductor to create a force of 5 N on the conductor.

$$F = B \times I \times l$$

$$5 = 0.35 \times I \times 0.2$$

Transpose for I

$$I = \frac{5}{(0.35 \times 0.2)} = 71.42\,\text{A (note use of brackets)}$$

Enter in calculator

$$5 \div (0.35 \times 0.2) =$$

Induced e.m.f.

When a conductor is moved through a magnetic field at right angles to it an e.m.f. is induced in the conductor.

$$e = Blv \text{ volts}$$

where B is the flux density (T), l is the length of conductor within the magnetic field (m), v is the velocity of the conductor (metres per second, m/s).

EXAMPLE I   Calculate the induced e.m.f. in a conductor with an effective length of 0.25 m moving at right angles, at a velocity of 5 m/s, through a magnetic field with a flux density of 1.6 tesla.

$$e = B \times l \times v$$

$$e = 1.6 \times 0.25 \times 5 = 2 \, \text{volts}$$

EXAMPLE 2   The e.m.f. in a conductor of effective length 0.25 m moving at right angles through a magnetic field at a velocity of 5 m/s is 1.375 V. Calculate the magnetic flux density.

$$E = Blv$$

(**Note** E is volts in this equation not V)

$$1.375 = B \times 0.25 \times 5$$

Transpose

$$B = \frac{1.375}{(0.25 \times 5)} = 1.1 \, \text{T}$$

Self-inductance

If the self-inductance of a magnetic system is L henrys and the current changes from I_1 at time t_1 to I_2 at time t_2, the induced e.m.f. is

$$e = L \times \text{ rate of change of current}$$

$$= L \times \frac{I_2 - I_1}{t_2 - t_1} \text{ volts}$$

where the current is in amperes and the time in seconds.

EXAMPLE 1   A coil has self-inductance $3\,\text{H}$, and the current through it changes from $0.5\,\text{A}$ to $0.1\,\text{A}$ in $0.01\,\text{s}$. Calculate the e.m.f. induced.

$$e = L \times \text{ rate of change of current}$$

$$= 3 \times \frac{0.5 - 0.1}{0.01}$$

$$= 120 \text{ V}$$

The self-inductance of a magnetic circuit is given by

$$\text{self-inductance} = \frac{\text{change in flux linkage}}{\text{corresponding change in current}}$$

$$L = N \times \frac{\Phi_2 - \Phi_1}{I_2 - I_1} \text{ henrys}$$

where N is the number of turns on the magnetizing coil and Φ_2, I_2; Φ_1, I_1 are corresponding values of flux and current.

EXAMPLE 2   The four field coils of a d.c. machine each have 1250 turns and are connected in series. The change in flux

produced by a change in current of 0.25 A is 0.0035 Wb. Calculate
the self-inductance of the system.

$$L = N \times \frac{\Phi_2 - \Phi_1}{I_2 - I_1}$$

$$= 4 \times 1250 \times \frac{0.0035}{0.25}$$

$$= 70 \text{ H}$$

Mutual inductance

If two coils A and B have mutual inductance M henrys, the e.m.f. in coil A due to current change in coil B is

$$e_A = M \times \text{rate of change of current in coil B}$$

Thus, if the current in coil B has values I_1 and I_2 at instants of time t_1 and t_2,

$$e = M \times \frac{I_2 - I_1}{t_2 - t_1} \text{ volts}$$

EXAMPLE 1   Two coils have mutual inductance 3 H. If the current through one coil changes from 0.1 A to 0.4 A in 0.15 s, calculate the e.m.f. induced in the other coil.

$$e = 3 \times \frac{0.4 \times 0.1}{0.15} \quad (t_2 - t_1 = 0.15)$$

$$= 6 \text{ V}$$

The mutual inductance between two coils is given by

$$M = N_A \times \frac{\Phi_2 - \Phi_1}{I_{B1} - I_{B2}} \text{ henrys}$$

where N_A is the number of turns on coil A and Φ_2 and Φ_1 are the values of flux linking coil A due to the two values of current in coil B, I_{B2} and I_{B1}, respectively.

EXAMPLE 2   The secondary winding of a transformer has 200 turns. When the primary current is 1 A the total flux is 0.05 Wb, and when it is 2 A the total flux is 0.095 Wb. Assuming

that all the flux links both windings, calculate the mutual
inductance between the primary and secondary.

$$M = N_A \times \frac{\Phi_2 - \Phi_1}{I_{B1} - I_{B2}}$$

$$= 200 \times \frac{0.095 - 0.05}{2 - 1}$$

$$= 9\,H$$

EXERCISE 13

1. Convert (a) 0.001 25 Wb to milliwebers, (b) 795 000 μWb to
   webers.
2. Complete the following table:

Wb	0.025		0.74	
mWb		35		
μWb			59 500	850 000

3. The flux density in an air gap of cross-sectional area
   $0.0625\,m^2$ is I.I.T. Calculate the total flux.
4. Determine the flux density in an air gap 120 mm by 80 mm
   when the total flux is 7.68 mWb.
5. An air gap is of circular cross-section 40 mm in diameter.
   Find the total flux when the flux density is 0.75 T.
6. Calculate the force on a conductor 150 mm long situated at
   right angles to a magnetic field of flux density 0.85 T and
   carrying a current of 15 A.
7. Determine the flux density in a magnetic field in which a
   conductor 0.3 m long situated at right angles and carrying
   a current of 15 A experiences a force of 3.5 N.
8. Complete the table below, which relates to the force on
   conductors in magnetic fields:

Flux density (T)	0.95	0.296	1.2	0.56	
Conductor length (m)	0.035		0.3	0.071	0.5
Current (A)		4.5		0.5	85
Force (N)	0.05	0.16	12		30

9. A conductor 250 mm long is situated at right angles to a
   magnetic field of flux density 0.8 T. Choose six values of

current from 0 to 5 A, calculate the force produced in each case, and plot a graph showing the relationship between force and current.

10. If the conductor of question 9 is to experience a constant force of 1.5 N with six values of flux density ranging from 0.5 T to 1.0 T, calculate the current required in each case and plot a graph showing the relationship between current and flux density.

11. A conductor 250 mm long traverses a magnetic field of flux density 0.8 T at right angles. Choose six values of velocity from 5 to 10 m/s. Calculate the induced e.m.f. in each case and plot a graph of e.m.f. against velocity.

12. If the conductor of question 11 is to experience a constant induced e.m.f. of 3 V with values of flux density varying from 0.5 T to 1.0 T, choose six values of flux density, calculate in each case the velocity required, and plot a graph of velocity against flux density.

13. A conductor of effective length 0.2 m moves through a uniform magnetic field of density 0.8 T with a velocity of 0.5 m/s. Calculate the e.m.f. induced in the conductor.

14. Calculate the velocity with which a conductor 0.3 m long must pass at right angles through a magnetic field of flux density 0.65 T in order that the induced e.m.f. shall be 0.5 V.

15. Calculate the e.m.f. induced in a coil of 1200 turns when the flux linking with it changes from 0.03 Wb to 0.045 Wb in 0.1 s.

16. The magnetic flux in a coil of 850 turns is 0.015 Wb. Calculate the e.m.f. induced when this flux is reversed in 0.25 s.

17. A coil has self-inductance 0.65 H. Calculate the e.m.f. induced in the coil when the current through it changes at the rate of 10 A/s.

18. A current of 5 A through a certain coil is reversed in 0.1 s, and the induced e.m.f. is 15 V. Calculate the self-inductance of the coil.

19. A coil has 2000 turns. When the current through the coil is 0.5 A the flux is 0.03 Wb; when the current is 0.8 A the flux is 0.045 Wb. Calculate the self-inductance of the coil.

20. An air-cored coil has 250 turns. The flux produced by a current of 5 A is 0.035 Wb. Calculate the self-inductance of the coil. (Hint: in an air-cored coil, current and magnetic flux are directly proportional. When there is no current, there is no flux.)

21. Two coils have mutual inductance 2 H. Calculate the e.m.f. induced in one coil when the current through the other changes at the rate of 25 A/s.

22. Two coils have mutual inductance 0.15 H. At what rate must the current through one change in order to induce an e.m.f. of 10 V in the other?

23. Two coils are arranged so that the same flux links both. One coil has 1200 turns. When the current through the other coil is 1.5 A, the flux is 0.045 Wb; when the current is 2.5 A the flux is 0.07 Wb. Calculate the mutual inductance between the coils.

24. Calculate the e.m.f. induced in one of the coils of question 23 if a current of 0.2 A in the other coil is reversed in 0.15 s.

25. The total magnetic flux in an air gap is given as 200 μW. In milliwebers this is
    (a) 0.2   (b) 20   (c) 0.02   (d) 2

26. The total flux in a magnetic circuit is 0.375 mWb and the cross-sectional area is 5 cm^2. The flux density is
    (a) 1.333 T   (b) 0.075 T   (c) 0.75 T   (d) 7.5 T

27. A force of 0.16 N is experienced by a conductor 500 mm long carrying a current of 0.375 A and resting at right angles to a uniform magnetic field. The magnetic flux density is
    (a) 0.117 T   (b) 0.85 T   (c) 8.5 T   (d) 0.085 T

28. The e.m.f. induced in a conductor of length 0.15 m moving at right angles to a magnetic field with a velocity of 7.5 m/s is 22.5 mV. The magnetic flux density is
    (a) 20 T   (b) 25.3 T   (c) 0.02 T   (d) 0.0253 T

29. The magnetic flux linking a coil of 150 turns changes from 0.05 Wb to 0.075 Wb in 5 ms. The e.m.f. induced is

    (a) 750 V   **(b)** 0.75 V   **(c)** 37.5 V   **(d)** 37 500 V

30. When the current through a coil changes from 0.15 A to 0.7 A in 0.015 s, the e.m.f. induced is 100 V. The self-inductance of the coil is

    (a) 367 H   **(b)** 0.367 H   **(c)** 2.73 H   **(d)** 1.76 H

31. Two coils have mutual inductance 0.12 H. The current through one coil changes at the rate of 150 A/s. The e.m.f. induced in the other is

    (a) 1250 V   **(b)** 0.0008 V   **(c)** 180 V   **(d)** 18 V

Cable selection

When a current is passed through a conductor it causes it to rise in temperature.

HEAT IN CABLES

When installing circuits it is important that the correct size current carrying conductor is selected to carry the current required without causing the cable to overheat and that the voltage drop caused by the resistance of the cable is not greater than is permissible.

The following calculations are designed to compensate for conductor temperature rise.

- We must first calculate the design current that the circuit will have to carry (I_b).

- Calculation is

$$\frac{P}{V} = I$$

(I being design current).

- A protective device must now be selected (I_n) this must be equal to or greater than I_b.

- If the cable is to be installed in areas where environmental conditions will not allow the cable to cool, correction factors will be required.

- C_a is a factor to be used where ambient temperature is above or below 30°C. This factor can be found in Table 4C1, appendix 4 of BS 7671. If a BS 3036 rewirable fuse is needed Table 4C2 should be used.

- C_g is a factor to be used where the cable is grouped or bunched (touching) with other cables. This factor can be found in Table 4B1, appendix 4 of BS 7671.

- C_i is a factor for use where a conductor is surrounded by thermal insulation and can be found in Table 52A part 5 of BS 7671.

- C_r is a factor for rewirable fuses and is always 0.725. This factor must always be used when rewirable fuses protect a circuit. The reason for the factor will be explained at end of the chapter.

- These factors should be multiplied together and then divided into I_n.

- Therefore the calculation is

$$\frac{I_n}{C_a \times C_g \times C_i \times C_r}$$

- The current carrying capacity of the cable must be equal to or greater than the result of this calculation.

- It should be remembered that only the correction factors that effect the cable at the same time should be used.

EXAMPLE   A circuit is to be installed using 2.5 mm^2, 1.5 mm^2 twin and earth 70°C thermoplastic cables, it is 32 metres long and protected by a BS 88 fuse. The load to be supplied is a 4.2 kW kiln, the circuit is to be installed in minitrunking containing one other circuit at an ambient temperature of 35°C. Maximum permissible volt drop is 7 V. Supply is a T N S system with a Ze of 0.7 Ω. Calculate the minimum cable that may be used.
Design current

$$I_b = \frac{P}{V}$$

$$\frac{4.2 \times 1000}{230} = 18.26 \text{ amperes}$$

Protective device I_n (≥ 18.26), nearest BS 88 is 20 amperes.

In the example, the cable is installed in plastic trunking. From BS 7671 Table 4A1 Installation methods, number 8, method 3 matches the example.

The cable is installed in trunking which will contain one other circuit. Correction factor for grouping (C_a) is required from BS 7671 Table 4B1. It can be seen that for two circuits in one enclosure a factor of 0.8 must be used.

The ambient temperature is 35°C. A correction factor for ambient temperature (C_a) from Table 4C1 must be used. Thermoplastic cable at 35°C a factor of 0.94.

Using these factors, it is now possible to calculate the minimum size conductors required for this circuit.

$$I_t \geq \frac{I_n}{C_a \times C_g}$$

$$I_t \geq \frac{20}{0.8 \times 0.94} = 26.59 \text{ A}$$

Calculator method

$$20 \div (0.8 \times 0.94) = 26.59$$

This is the minimum value of current that the cable must be able to carry to enable it to be installed in the environmental conditions affecting the cable.

From Table 4D2A columns 1 and 4, it can be seen that a 4 mm^2 cable has an I_t (current carrying capacity) of 30 amperes.

A cable with 4 mm^2 live conductors will carry the current in these conditions without overheating, but will it comply with the voltage drop requirements?

From Table 4D2B columns 1 and 3, it can be seen that 4 mm^2 cable has a voltage drop of 11 (mV/A/m) or millivolts × load current × length of circuit. As the value is in millivolts, it must be converted to volts by dividing by 1000.

The circuit length is 32 metres and the load current is 18.26 amperes.

Calculation

$$\frac{mV/A/m \times I \times L}{1000} = \frac{11 \times 18.26 \times 32}{1000} = 6.42 \text{ V}$$

The voltage drop in this cable will be 6.42 V which is acceptable as the maximum permissible for the circuit is 7 V.

The calculations which have been carried out up to this point have been to select a cable to comply with the current and voltage drop requirements for the circuit. This is only part of the calculation. It is now important that a calculation is carried out to prove that the protective device will operate within the time required if an earth fault were to occur on the circuit.

The load is classed as fixed equipment, this will have a disconnection time not exceeding 5 seconds (regulation 413-02-13).

The resistance of the cable must now be calculated:
A 4 mm^2 twin and earth cable will have a circuit protective conductor (CPC) of 1.5mm^2.

From Table 9A in the *On-Site Guide*, it can be seen that this cable will have a resistance of 16.71 milli-ohms per metre at 20°C.

As the cable could operate at 70°C the multiplier from Table 9C in the *On-Site Guide* must be used to adjust the resistance value from 20°C to 70°C.

Calculation

M Ω × length × multiplier 1.2 ÷ 1000 (to convert to ohms)

$$\frac{16.71 \times 32 \times 1.2}{1000} = 0.64 \text{ Ohms}$$

The resistance of the cable at operating temperature of 70°C is 0.64 Ω.

Z_s (earth loop impedance) must now be calculated.

$$Z_s = Z_e + (r_1 + r_2)$$

From the information given in the example, Z_e (external earth loop impedance) is 0.7 Ω. Therefore,

$$Z_s = 0.7 + 0.64$$

$$Z_s = 1.34 \ \Omega$$

This value must now be checked against the value for maximum permissible Z_s. This is in BS 7671 Table 41D for 5 second disconnection. It can be seen that the maximum Z_s for a 20 A BS 88 fuse is 3.04 Ω. As the circuit has a calculated Z_s of 1.34, this will be satisfactory.

EXERCISE 14

1. The voltage drop figure for a given cable is 44 mV/A/m. Calculate the voltage drop in a 15 m run of this cable when carrying a load of 6 A.
2. The design current of a circuit protected by a BS 1361 fuse is 28 A, the grouping correction factor is 0.8, and the ambient temperature correction factor is 1.04. Calculate the minimum current-carrying capacity of the cable.
3. A circuit is protected by a BS 3871 circuit breaker rated at 30 A. The grouping correction factor is 0.54 and the ambient temperature correction factor is 0.94. Calculate the minimum current capacity of the cable.
4. Calculate the effect on the minimum cable current rating required in question 3 if the circuit breaker is replaced by a BS 3036 semi-enclosed fuse.
5. A cable with a voltage drop figure of 6.4 mV/A/m supplies a current of 24 A to a point 18 m away from a 230 V supply source. Determine (a) the voltage drop in the cable and (b) the actual voltage at the load point.
6. There is a voltage drop limitation of 5 V for a circuit wired in pvc-insulated twin and earth cable (clipped direct), having a length of run of 35 m. The current demand is assessed as 36 A. Protection is by a BS 1361 fuse. Establish the

   (a) fuse rating,
   (b) maximum mV/A/m value,
   (c) minimum cable cross-sectional area.

DISCONNECTION TIMES FOR FUSES

BS 7671 requirements, Part 4, Chapters 41 and 47, give maximum disconnection times for circuits under earth fault conditions.

Maximum disconnection time for 230 V circuits feeding only stationary equipment is 5 seconds (regulation 413-02-13). Table 41D gives maximum total earth fault loop values (Z_s) permissible to achieve this disconnection time.

Maximum disconnection times for 230 V socket outlet circuits and other 230 V circuits that supply portable equipment should not exceed 0.4 seconds (regulation 413-02-09 and Table 41A). Table 41B1 gives the maximum earth fault loop impedance, Z_s to achieve this disconnection time.

Maximum disconnection time for 230 V circuits feeding only stationary equipment is 5 seconds (regulation 413-02-13). Table 41D gives guidance to the maximum total earth fault loop impedance values (Z_s) to achieve this disconnection time.

EXAMPLE   A circuit protected by a BS 3036 fuse is feeding stationary equipment. If Table 41D is consulted, this states that the maximum value of Z_s for a 30 A BS 3036 fuse should not exceed 2.76 Ω.

Maximum disconnection times for 230 V socket outlet circuits and other 230 V final circuits which supply portable equipment should not exceed 0.4 s (regulation 413-02-09 and Table 41A).

If a circuit supplying hand-held equipment was protected by a 30 A BS 3036 fuse, the Z_s value can be found in Table 41B. The maximum permissible Z_s for this circuit is now 1.14 Ω (considerably less than for a 5 second disconnection time).

Tables 41B1 and 41D should be used to establish the maximum earth fault loop impedance (Z_s) to achieve a 0.4 or 5 s disconnection time.

DISCONNECTION TIMES FOR CIRCUIT BREAKERS

When circuit breakers are used it is important that the maximum Z_s values from Table 41B2 are used. When these values are used, compliance with regulations 413-02-13 and 413-02-09 will be achieved as these devices are constructed to operate within 0.1 second, providing the correct value of maximum Z_s is used.

BS 7671 requirements, Part 4, Chapters 41 and 47 lay down maximum disconnection times for circuits under earth fault conditions.

Circuit breakers to BS EN 60898 are available as three types: B, C and D. It is important that the correct type is selected.

Type B will operate in 0.1 s when a maximum current of 5 times its current rating passes through it.
Type C will operate in 0.1 s when a maximum current of 10 times its current rating passes through it.
Type D will operate in 0.1 s when a maximum current of 20 times its current rating passes through it.

To allow this amount of current to flow, the resistance of the circuit $(R_1 + R_2)$ must be low enough. For circuit breakers, the maximum permissible Z_s can be calculated if required using Ohm's law.

EXAMPLE   A 20 amp BS EN 60898 device must operate at a maximum of 5 times its rating.

$$5 \times 20 = 100 \text{ A}$$

If this current value is now divided into the open circuit voltage, U_{Oc} the Z_s for the circuit will be calculated.

$$\frac{240}{100} = 2.4 \ \Omega$$

This is the maximum Z_s for a 20 amp type B device.
For a 20 amp type C device

$$10 \times 20 = 200 \text{ A}$$

$$\frac{240}{200} = 1.2 \ \Omega$$

For a 20 amp type D device

$$20 \times 20 = 400\,\text{A}$$

$$\frac{240}{400} = 0.6\,\Omega$$

This calculation can be used to calculate the Z_s for any
BS EN 60898 device.

The overload characteristic for these devices is the same for each
type, i.e. they will all operate at a maximum of 1.45 times their
current rating.

FUSING FACTORS, OVERLOAD AND FAULT CURRENT

Fuse factor (I_2)

Regulation 433-02-01 describes the characteristics required of
fuses to comply with BS 7671.

The current causing effective operation for overload (I_2) of a
protective device must not be greater than 1.45 times the current
carrying capacity of the conductor that it is protecting.

Apart from BS 3036 semi-enclosed fuses, all other protective
devices are manufactured to comply with regulation 433-02-01.

A BS 3036 semi-enclosed fuse will not operate on overload until
the current passing through it reaches approximately twice its
rating.

A fusing factor of 0.725 must be used when using BS 3036 fuses.

EXAMPLE A circuit is required to carry a load of 14 amperes.
The protective device must be $\geq 14\,\text{A}$. The nearest rating BS 3036
fuse is 15 A. The cable selected for this circuit must be calculated
using the following calculation.

Cable rating must be

$$\frac{15}{0.725} = 20.69\,\text{A minimum}$$

The BS 3036 fuse will operate at $15 \times 2 = 30$

The minimum cable rating of $20.69\,\text{A} \times 1.45 = 30\,\text{A}$ will satisfy regulation 433-02-01.

This factor is to be used on all circuits using BS 3036 for overload protection and must be used with any other correction factors for circuits as described in the chapter for cable selection.

SHORT CIRCUIT CURRENT

Is a current which will flow in a circuit of negligible impedance between live conductors.

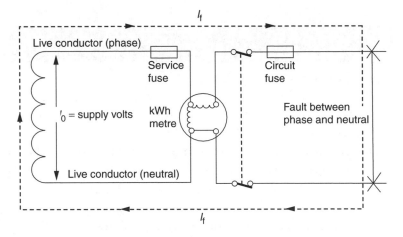

Fig. 42

Figure 42 shows the supply phase and neutral each having a resistance of $0.02\,\Omega$ and the final circuit phase and neutral each having a resistance of $0.23\,\Omega$. The total resistance of the supply and final circuit will be:

$$0.02 + 0.02 + 0.11 + 0.11 = 0.26\,\Omega$$

To calculate the short circuit current, the open circuit current, U_{Oc} of the supply transformer should be used. From appendix 3 of BS 7671 it can be seen that this is 240 V

$$\frac{240}{0.26} = 923\,\text{A}$$

Earth fault loop impedance

EARTH FAULT LOOP IMPEDANCE, Z_e

Z_e is the external earth fault loop impedance (resistance) phase conductor and earthing arrangement of the supply (see Figure 43).

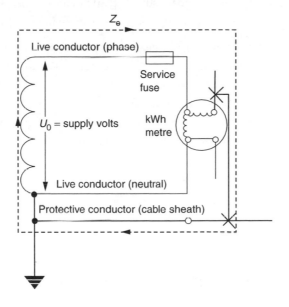

Fig. 43

Impedance of phase conductor 0.02 Ω

Impedance of earth path 0.78 Ω

Total external earth loop impedance

$$Z_e = 0.02 + 0.78 = 0.8 \ \Omega$$

Earth fault current if measured at the ends of the supply cable (origin) can be calculated:

$$\frac{240}{0.8} = 300\,\text{A}$$

It is important to use 240 V in this calculation as it is the open circuit voltage of the supply transformer.

EXAMPLE The 100 A high breaking current service fuse (BS 1361) at the origin of an installation has a fusing factor of 1.4, the open circuit voltage of the supply transformer (U_{Oc}) is 240 V, and the tested value of Z_e at the origin of the installation is 0.38 Ω.

(a) Calculate the minimum current required to blow the fuse.
(b) How much current will flow if the phase conductor comes into contact with the earthed sheath of the supply cable at the origin of the installation?
(c) Using appendix 3 from BS 7671 state approximately the time in which the current must operate under conditions in (b)?
Minimum fusing current

$$(I_2) = 100 \times 1.4$$

$$= 140\,\text{A}$$

Earth fault current

$$(I_f) = \frac{U_{Oc}}{Z_e}$$

$$= \frac{240}{0.38}$$

$$= 631.6\,\text{A}$$

(d) Using the table attached to Figure 3.1 from BS 7671, it can be seen that for a 5 second disconnection time a minimum current of 630 A is required to operate the fuse. (The higher the current the quicker the disconnection time.)

EARTH FAULT LOOP IMPEDANCE, Z_s

Z_s is the total earth fault loop impedance of the supply and the resistance of the final circuit cables, phase conductor (R_1) and circuit protective conductor (R_2)

$$Z_s = Z_e + R_1 + R_2$$

If the loop impedance of a system (Z_s) is high, the fault current will be low and the device protecting the circuit may not operate within the required time, this will result in the extraneous and exposed conductive parts within the circuit rising in potential and becoming a serious shock risk.

EXAMPLE A circuit is to be wired in $70\,^\circ$C thermoplastic cable with copper 2.5 mm^2 phase and 1.5 mm^2 circuit protective conductors. The circuit is 30 metres long and the Z_e for the circuit is measured at $0.35\,\Omega$.

(a) Calculate Z_s

(b) Calculate earth fault current

From Table 9A in the *On-Site Guide*, it can be found that a 2.5/1.5 mm^2 cable with copper conductors will have a ($r_1 + r_2$) value of 19.41 mΩ/m.

See chapter on voltage drop for use of 1.2 multiplier to correct operating resistance from $20\,^\circ$C to $70\,^\circ$C (operating temperature).

(a) Total resistance of final circuit cables

$$\frac{19.41 \times 30 \times 1.2}{1000} = 0.698$$

$$Z_s = Z_e + R_1 + R_2$$

$$Z_s = 0.35 + 0.698$$

$$Z_s = 1.04$$

(b) Earth fault current $\dfrac{240}{1.04} = 230.76\,\text{A}$

1. Complete the following table:

U (volts)	10	20		40	
I (amperes)	1		3	4	5
R (ohms)		10	10		10

2. Using the values from question 1, correctly completed, plot a graph of current against voltage. Take $1\ \text{cm} \equiv 1\,\text{A}$ vertically and $1\ \text{cm} \equiv 10\,\text{V}$ horizontally.

   Read from the graph the voltage required to produce a current of 3.6 A.

3. Complete the following table:

U (volts)		240		240	
I (amperes)	12	6	4	3	2.4
R (ohms)	20		60		100

4. When the table of question 3 has been correctly completed, plot a graph showing the relationship between current and resistance. Use the graph to find the value of the current when the resistance is $78\,\Omega$. State also the value of resistance required to give a current of 9.5 A.

5. Complete the following table:

U (volts)	100	100		56	96	132	84	144		
I (amperes)	10		12	7	8	12		12	11	9
R (ohms)		10	8				12		11	7

6. Complete the following table:

I (amperes)	100		10		0.1	0.1		0.001	0.1	
R (ohms)	0.1	1000	0.1	1000	0.1		0.1			0.01
U (volts)		100		10		100	10	20	200	2

7. Complete the following table:

R (ohms)		14		16		0.07	12		0.75	15
I (amperes)	0.5	15	0.05		1.2	0.9		0.2		
U (volts)	240		25	96	132		8.4	100	6	120

8. A cable of resistance $0.029\,\Omega$ carries a current of 83 A. What will be the voltage drop?

9. To comply with BS 7671 regulations, the maximum value of voltage drop which can be tolerated in a circuit supplied

from the 230 V mains is 9.2 V. Calculate the maximum resistance which can be allowed for circuits carrying the following currents:

(a) 28 A   (c) 77 A   (e) 203 A

(b) 53 A   (d) 13 A

10. The cable in a circuit has a resistance of 0.528 Ω. What is the maximum current it can carry if the voltage drop is not to exceed 5.8 V?

11. A 50 V a.c. system supplies the following loads by means of a radial circuit:

Load A: 15 A at a distance of 18 m from the supply point S,

Load B: 25 A at a distance of 35 m from the supply point A,

Load C: 20 A at a distance of 43 m from the supply point B.

The type of cable used produces a voltage drop of 2.7 mV per ampere per metre.

Calculate the voltage drop in each section of the circuit and the voltage at each load point.

12. Assuming a fusing factor of 1.4, complete the following table, which refers to various sizes of fuse:

Nominal current (A)   5   15   30   60   100

Minimum fusing

current (A)

13. Repeat exercise 12 using a fusing factor of 1.2.

14. Calculate the voltage between the metal parts and earth under the following fault conditions. The supply voltage in each case is 230 V.

	Rating of circuit fuse (I_n) (A)	Fusing factor	Resistance of CPC (R_2) etc. (Ω)	Resistance of remainder of loop $(Z_e + R_1)$ (Ω)
(a)	30	1.3	5	0.5
(b)	30	1.2	8	0.5
(c)	30	1.2	0.5	7
(d)	60	1.5	4	1
(e)	100	1.2	4	1
(f)	100	1.2	2.5	0.5

15. A current of 1.5 A flows in a 25 Ω resistor. The voltage drop is:

    (a) 0.06 V   (b) 37.5 V   (c) 16.67 V   (d) 3.75 V

16. If a cable must carry a current of 19.5 A with a voltage drop of not more than 6 V, its resistance must not exceed.

    (a) 32.5 Ω   (b) 117 Ω   (c) 0.308 Ω   (d) 3.25 Ω

17. A fuse rated at 30 A has a fusing factor of 1.4. The current required to blow the fuse is

    (a) 31.4 A   (b) 21.4 A   (c) 42 A   (d) 30 A

18. A faulty earthing conductor has a resistance of 12.5 Ω, and the resistance of the remainder of the fault path is 1.5 Ω. The supply voltage is 230 V. The voltage appearing between metal parts and earth is

    (a) 205.4 V   (b) 238.5 V   (c) 24.6 V   (d) 217.7 V

Materials costs, discounts and VAT

To find the value added tax due on an item:

EXAMPLE 1   If a consumer's unit with a main switch was quoted by the suppliers as costing £53.85 plus VAT. Calculate the cost of the item including VAT

$$\text{VAT on item} = \frac{53.85 \times 17.5}{100} = 9.42$$

Item including VAT £53.85 + £9.42 = £63.27

or enter on calculator 53.85 × 17.5% (9.423) + (63.27)

   (Figures in brackets are the answers that you will get.

   Do not enter into calculator.)

A difficulty that often occurs is that a value is given including VAT and it is necessary to calculate the item cost without the VAT.

EXAMPLE 2   Using the values from example 1:
A consumer's unit costs £63.27 inclusive of VAT. Calculate the VAT content.

$$\frac{63.27 \times 100}{117} = 53.86 \quad \text{OR} \quad \frac{63.27}{1.175} = 53.85$$

therefore the VAT content is £63.27 − £53.85 = £9.42 or multiply by 7 and divide by 47 to find VAT content

Answer £9.42

Enter on calculator $63.27 \times 7 \div 47 =$

or transpose formula from example 1 to find cost of unit without VAT

$$(53.85) \times 1.175 = 63.27$$

Transpose $(53.85) = \dfrac{63.27}{1.175}$

Cost less VAT is £53.85.

EXAMPLE 3 A 100 metres of 4 mm^2 3-core steel wired armoured cable costs £258.60 per 100 metres.

If a trade discount of 30% were allowed on this cable, calculate the cost of 60 metres.

$$30\% \text{ of } £258.60 = \frac{30 \times 258.60}{100} = £77.58$$

Trade price of cable per 100 metres

£258.60 − £77.58 = £181.02

Cost of 1 metre of this cable at trade price is

$$\frac{181.02}{100} = 1.81$$

60 metres would cost £1.81 × 60 m = £108.60

A far easier method would be to use a calculator and enter:

$$258.6 \times 30\% = (77.58) - = (181.02) \div 100 \times 60 = 108.60$$

Figures in brackets are the answers that you will get.

Do not enter them into the calculator.

If value added tax needs to be added (current VAT rate is 17.5%)

$$\frac{108.6 \times 17.5}{100} = 19.00$$

£19.00 is the VAT on the cable and should be added to the trade cost

$$£108.6 + £19 = £127.60$$

Calculator method:
Enter $108.6 \times 17.5\% + (127.60)$ (answer in brackets)

1. If 60 lengths of cable tray cost £732 including VAT, calculate (a) the cost of each length, (b) the cost of 17 lengths.
2. If 66 m of black-enamelled heavy-gauge conduit cost £87 including VAT, calculate (a) the cost per metre, (b) the cost of 245 m.
3. If 400 woodscrews cost £4.52, calculate (a) the cost of 250 screws, (b) the number of screws which could be purchased for £5.28.
4. If 100 m of heavy-gauge plastic conduit is listed at £85.20, plus VAT at 17.5%, calculate the price of 100 m to the customer.
5. The list price of electrician's solder is £360 for 20 kg plus VAT at 17.5% and with a special trade discount of 25%. Calculate the invoice price of 4 kg of solder.
6. An invoice was made out for 20 lengths of 50 mm × 50 mm cable trunking. Each length cost £17.55 plus 17.5% VAT, less 35% trade discount. Calculate the invoice total.
7. An alteration to an existing installation requires the following material:
   12 m of plastic trunking at £6.23 per m,
   14.5 m of plastic conduit at £86 per 100 m,
   45 m of cable at £14.60 per 100 m,
   29 single socket-outlets at £12.15 each,
   saddles, screws, plugs, etc. £9.20.
   Calculate the total cost of the materials. VAT is chargeable at 17.5%.

8. An order was placed one year ago for the following items:
   135 m MIMS cable at £217 per 100 m,
   500 pot-type seals/glands at £26 per 10,
   200 one-hole clips at £29 per 100.
   Calculate (a) the original cost of this order; (b) the present cost of the order, allowing 15% per annum for inflation. VAT is chargeable at 17.5% at both (a) and (b).

9. The materials list for an installation is as follows:
   45 m of 1.00 mm^2 twin with earth cable at £19.30 per 100 m,
   45 m of 2.5 mm^2 twin with earth cable at £28.20 per 100 m,
   nine two-gang one-way switches at £3.50 each,
   two two-way switches at £2.12 each,
   six single switched socket outlets at £3.35 each,
   two twin switched socket outlets at £6.40 each,
   one consumer unit at £62.20,
   sheathing, screws, plugs, etc. £8.00.
   Determine the total cost of the materials for this work, adding 17.5% VAT.

10. A contractor's order for conduit and fittings reads as follows:
    360 m of 20 mm BEHG steel conduit at £147 per 90 m,
    50 20 mm BE standard circular terminal end boxes at £1.81 each,
    50 20 mm BE standard through boxes at £2.17 each,
    50 20 mm BE standard tee boxes at £2.57 each,
    50 20 mm spacer-bar saddles at £23.20 per 100,
    50 20 mm steel locknuts at £14.90 per 100,
    50 20 mm brass hexagon male bushes at £38 per 100.
    All prices are list, the contractor's discount on all items is 40%, and VAT is chargeable at 17.5%. Calculate the invoice total for this order.

11. It is necessary to install six tungsten-halogen flood lighting luminaires outside a factory and the following equipment is required.
    Manufacturer's list prices are as shown:

6 off 500 W 'Teck' T/H luminaires	at £14.50 each*,
1 off 'Teck' PIR sensor/relay unit	at £24.10 each*,
80 m 20 mm galvanized steel conduit	at £186 per 100 m,

6 off 20 mm galvanized tee boxes	at £275 per 100,
1 off 20 mm galvanized angle box	at £265 per 100,
7 off galvanized box lids and screws	at £11 per 100,
8 off 20 mm galvanized couplings	at £19 per 100,
30 off 20 mm spacer saddles	at £17.20 per 100,
14 off 20 mm brass male bushes	at £38 per 100,
1 off 'TYLOR' 20 A switch-fuse	at £24.50 each*,
1 off 'TYLOR' 10 A one-way switch	at £3.20 each*,
180 m 1.5 mm^2 pvc single cable	at £12.15 per 100 m,
3 m 0.75 mm^2 three-core pvc flex	at £26.30 per 100 m,
9 off 10 A three-way porcelain connectors	at £80 per 100.

Plus sundries taken from own stock, allow £15.

The wholesaler offers a 25% discount on non-branded items and 10% on branded * items. Calculate (a) the basic cost of the materials and (b) the total cost including VAT at 17.5%.

12. The list prices of certain equipment are as follows:
    (a) £570.30 with 25% discount,
    (b) £886.20 with 40% discount,
    (c) £1357.40 with 10% discount,
    (d) £96.73 with 35% discount.

    For each of the above establish:
    (i) the basic cost price,
    (ii) the VAT chargeable.

13. For each of the following VAT inclusive prices establish the basic cost price:
    (a) £656.25   (d) £1025.27
    (b) £735.33   (e) £3257.72
    (c) £895.43

14. A certain cable is priced at £19.50 per 100 m plus 17.5% VAT. The cost of 65 m is:
    (a) £22.91   (b) £16.09   (c) £14.89   (d) £10.46

15. A certain item of equipment was invoiced at £25.75 and this included VAT at 17.5%. The list price of the item was:
    (a) £3.84   (b) £21.91   (c) £30.26   (d) £43.25

Electrostatics

THE PARALLEL PLATE CAPACITOR

When a capacitor is connected to a d.c. supply it becomes charged, the quantity of charge is in coulombs.

$$Q = CU$$

where Q = quantity, C = capacitance in farads and U = voltage.

EXAMPLE 1   A 70 µF capacitor is connected to a 150 volt d.c. supply. Calculate the charge stored in the capacitor.

$$Q = C \times U$$
$$= 70 \times 10^{-6} \times 150$$
$$= 0.0105 \text{ coulombs}$$

Enter into calculator $70 \times \text{EXP} - 6 \times 150 =$

Energy stored is in watts or Joules.

$$W = \frac{1}{2} CU^2$$

EXAMPLE 2   Calculate the energy stored in a 120 µF capacitor when connected to a 110 volt d.c. supply.

$$W = \frac{120 \times 10^{-6} \times 110}{2}$$
$$= 6.6^{-03} \text{ or } 0.006 \text{ Joules}$$

Enter into calculator $120 \text{ EXP} - 6 \times 110 \div 2 =$

SERIES ARRANGEMENT OF CAPACITORS

If a number of capacitors are connected in series, the total capacitance can be calculated.

$$\frac{1}{C_1} + \frac{1}{C_2} + \frac{1}{C_3} = \frac{1}{Ct} = C$$

The result will be as equivalent to a single capacitor.

EXAMPLE 1 Calculate the value of capacitance when capacitors of 23, 42 and 36 µF are connected in series.

$$\frac{1}{C} + \frac{1}{C} + \frac{1}{C} = \frac{1}{C} = C$$

$$= \frac{1}{23} + \frac{1}{42} + \frac{1}{36} = \frac{1}{C}$$

$$= 10.51 \ \mu F$$

Enter on calculator $23 \ X^{-1} + 42 \ X^{-1} + 36 \ X^{-1} = X^{-1} =$

EXAMPLE 2 Calculate the value of a capacitor which when connected in series with another of 20 µF will give a resulting capacitance of 12 µF.

$$\frac{1}{C} = \frac{1}{C} + \frac{1}{C}$$

$$\frac{1}{12} = \frac{1}{20} + \frac{1}{C}$$

$$\frac{1}{C} = \frac{1}{12} - \frac{1}{20}$$

$$= 30 \ \mu F$$

Enter on calculator $12 \ X^- - 20 \ X^- = X^- =$

EXAMPLE 3 Capacitors of 4, 6 and 12 µF are connected in series to a 300 volt d.c. supply. Calculate

(a) the total capacitance

(b) the charge stored

(c) the energy stored.

(a) $\dfrac{1}{C} = \dfrac{1}{4} + \dfrac{1}{6} + \dfrac{1}{12}$

$\qquad = \dfrac{1}{0.5}$

$\qquad = 2\,\mu F$

(b) Charge stored $Q = CU$

$\qquad\qquad = 2 \times 300$

$\qquad\qquad = 600\ \mu F$

(c) Energy stored $W = \dfrac{1}{2}CU^2$

$\qquad\qquad = \dfrac{600 \times 10^{-6} \times 300^2}{2}$ or

$\qquad\qquad 600 \times 10^{-6} \times 300^2 \times 0.5$

$\qquad\qquad = 27\,\text{Joules}$

PARALLEL ARRANGEMENT OF CAPACITORS

When a number of capacitors are connected in parallel they are equivalent to a single capacitor of value C given by

$$C = C_1 + C_2 + C_3,\ \text{etc.}$$

When the arrangement is connected to a d.c. supply voltage, the total charge is the sum of the charges stored in each capacitor.

$$Q = Q_1 + Q_2 + Q_3$$

Q_1 is the charge on C_1, etc. The voltage is common to all capacitors.

EXAMPLE I   Capacitors of 8 and 10 µF are connected in parallel to a 20 V supply. Calculate the charge stored on each and the total energy.

Charge on 8 µF capacitor is

$$Q = 8 \times 20$$
$$= 160 \text{ µC}$$

Charge on 10 µF capacitor is

$$Q = 10 \times 20$$
$$= 200 \text{ µC (microcoulombs as } C \text{ is in µF)}$$

Total energy is:

$$W = \frac{1}{2}CU^2$$
$$= \frac{1}{2} \times 18 \times 20^2$$
$$= 3600 \text{ µJ (as } C \text{ is in microfarads)}$$

EXAMPLE 2   Calculate the value of a single capacitor equivalent to the arrangement of capacitors of 4 µF and 6 µF in parallel and a 12 µF capacitor in series with them.

Capacitance of parallel group is

$$C_1 + C_2 = C$$
$$= 4 + 6$$
$$= 10 \text{ µF}$$

Treated as a single capacitor, this value can now be used with the capacitor in series to calculate the total capacitance.

$$\frac{1}{C} = \frac{1}{10} + \frac{1}{12}$$
$$= 5.45 \text{ µF}$$

1. Complete the following table, which refers to a certain variable capacitor:

Applied volts (U)	50		25	80	45
Capacitance (µF)		0.3	0.4		0.8
Charge (µC)	10	18		48	

2. Capacitors of 3 µF and 5 µF are connected in series to a 240 V d.c. supply. Calculate

   (a) the resultant capacitance,

   (b) the charge on each capacitor,

   (c) the p.d. on each capacitor,

   (d) the energy stored in each capacitor.

3. Calculate the value of a single capacitor equivalent to three 24 µF capacitors connected in series. What would be the value of ten 24 µF capacitors connected in series?

4. What value of capacitor connected in series with one of 20 µF will produce a resultant capacitance of 15 µF?

5. Three capacitors, of values 8 µF, 12 µF, and 16 µF, respectively, are connected across a 240 V d.c. supply, (a) in series and (b) in parallel. For each case, calculate the resultant capacitance and also the potential difference across each capacitor.

6. Calculate the value of the single capacitor equivalent to the arrangement shown in Figure 44.

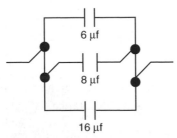

6 µf

8 µf

16 µf

Fig. 44

7. A 12 µF capacitor is charged to 25 V. The energy stored is
   (a) 150 µJ   (b) 3750 µJ   (c) 3750 J   (d) 150 J
8. Capacitors of 2 µF and 4 µF are connected in series. When an additional capacitor is connected in series, the combined capacitance falls to 1 µF. The value of the third capacitor is
   (a) 4 µF   (b) 0.5 µF   (c) 0.25 µF   (d) 1.2 µF
9. Capacitors of 8 µF, 12 µF, and 20 µF are connected in parallel. For a total charge of 4000 µC to be stored, the voltage to be applied to the combination is
   (a) 0.01 V   (c) 100 V
   (b) 15 480 V   (d) 1034 V

Formulae

Voltage	$U = I \times R$
Current	$I = \dfrac{U}{R}$
Resistance	$R = \dfrac{U}{I}$
Power	$P = U \times I$
Power loss	$P = I^2 R$
Current	$I = \dfrac{P}{U}$
Voltage	$U = \dfrac{P}{I}$
Resistors in parallel	$\dfrac{1}{R_1} + \dfrac{1}{R_2} + \dfrac{1}{R_3} = \dfrac{1}{R} \therefore R$
Area of a circle (mm^2 or m^2)	$\dfrac{\pi \times d^2}{4} = CSA$
Circumference of a circle (mm^2 or m^2)	$\pi \times d = C$
Area of triangle (mm^2 or m^2)	$\dfrac{1}{2} \text{ base } \times \text{ height}$
Resistance of a copper conductor (Ω)	$\dfrac{1.78 \times 10^{-8} \times L}{CSA \times 10^{-6}} = R$ *(where CSA is in mm^2)*

Resistance of an aluminium

conductor (Ω)

$$\frac{2.84 \times 10^{-8} \times L}{\text{CSA} \times 10^{-6}} = R$$

(where CSA *is in* mm^2*)*

Transformer calculation

$$\frac{Up}{Us} = \frac{Np}{Ns} = \frac{Is}{Ip}$$

Transformer efficiency

$$\frac{\text{power out}}{\text{power in}} = \text{per unit} \times 100 \text{ for } \%$$

WORK

$W = f \times d$

Work in N/m = force in Newtons $\times$ distance in mm or m

$1 \text{ kg} = 9.81$ Newtons

$$P = \frac{W}{t} \text{ or } \frac{\text{Workdone (Nm)}}{\text{Time (secs)}} = \text{Power in watts}$$

$J = W \times t$ or Energy (joules) = Watts $\times$ time in seconds

$$E = \frac{\text{Output}}{\text{Input}} \times 100 \text{ efficiency } in \%$$

CAPACITANCE

Charge of a capacitor is in coulombs $Q = CU$

Total charge of more than one capacitor $Q = Q_1 + Q_2 + Q_3$ etc.
or capacitance is

$$\frac{Q}{U} \text{ Farads}$$

Total capacitance of series connected

$$\frac{1}{C_1} + \frac{1}{C_2} + \frac{1}{C_3} \text{ etc.} = \frac{1}{C_T} = C$$

Total capacitance of parallel connected $C_1 + C_2 + C_3$ etc. $= C$

Energy stored in a capacitive circuit Energy $W = \frac{1}{2}CV^2$ Joules

Energy Stored in an inductive circuit

Energy $W = \frac{1}{2}LI^2$ Joules (where L is in henrys)

THREE-PHASE CALCULATIONS

I_p = phase current
I_L = line current
U_L = line voltage
U_P = phase voltage

In star (Only one current)

$$I_P = I_L$$

$$U_P = \frac{U_L}{\sqrt{3}}$$

$$U_L = U_P\sqrt{3}$$

$$P = \sqrt{3} \times U_L \times I_L$$

$$I_L = \frac{P}{\sqrt{3} \times U_L}$$

In circuits with power factor

$$P = \sqrt{3} \times U_L \times I_L \times \cos\phi$$

$$I_L = \frac{P}{\sqrt{3} \times U_L \times \cos\phi}$$

In Delta (only one voltage)

$$U_L = U_P$$

$$I_P = \frac{I_L}{\sqrt{3}}$$

$$I_L = I_P \times \sqrt{3}$$

$$P = \sqrt{3} \times U_L \times I_L$$

In circuits with power factor

$$P = \sqrt{3} \times U_L \times I_L \times \cos\phi$$

$$I_L = \frac{P}{\sqrt{3} \times U_L \times \cos\phi}$$

Power factor $\cos\phi = \dfrac{\text{True power}}{\text{Apparent power}} = \dfrac{\text{Watts}}{\text{Volt} \times \text{amps}}$

Pythagoras-type calculations

$$Z^2 = R^2 + X^2 \text{ or } Z = \sqrt{R^2 + X^2}$$

$$R^2 = Z^2 - X^2 \text{ or } R = \sqrt{Z^2 - X^2}$$

$$X^2 = Z^2 - R^2 \text{ or } X = \sqrt{Z^2 - R^2}$$

$$kVA^2 = kW^2 = kVAr^2 \text{ or } kVA = \sqrt{kW^2 + kVAr^2}$$

$$kW^2 = kVA^2 - kVAr^2 \text{ or } kW = \sqrt{kVA^2 - kVAr^2}$$

$$kVAr^2 = kVA^2 - kW^2 \text{ or } kVAr = \sqrt{kVA^2 - kW^2}$$

Capacitive reactance

$$X_C = \frac{1}{2\pi f C \times 10^{-6}} \text{ or } \frac{1 \times 10^6}{2\pi f C}$$

$$C = \frac{1}{2\pi f X \times 10^{-6}} \text{ or } \frac{1 \times 10^6}{2\pi f X}$$

Inductive reactance

$$X_L = 2\pi f L$$

$$L = \frac{X_L}{2\pi f X}$$

Synchronous speed and slip calculations

N_S is synchronous speed in revs/sec or $\times$ 60 for revs/min

N_R is speed of rotor in revs/sec or $\times$ 60 for revs/min

f is frequency of supply

P is pairs of poles

Unit slip is shown as a decimal

Percentage slip is shown as %

Synchronous speed

$$N_S = \frac{f}{P} \text{ in revs per sec} \times 60 \text{ for rpm}$$

Rotor speed

$$\frac{N_S - N_R}{N_S} = \text{unit slip} \times 100 \text{ for \%}$$

Calculations associated with cable selection

$$I_t \geq \frac{I_N}{\text{Correction factors}}$$

Cable resistance at 20°C

$$R = \frac{r_1 + r_2 \times \text{length in m}\Omega}{1000}$$

Volt drop in cable

$$\frac{\text{mV} \times \text{amperes} \times \text{length}}{1000}$$

Earth fault loop impedance $Z_s = Z_e = (R_1 + R_2)$

Electronic symbols

BS 3939 graphical symbols used in eletronics.
   Diagram No.

   1. AC relay.

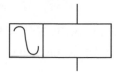

   Fig. I

   2. This is a symbol for a battery, the long line represents the positive terminal. Each pair of lines is one cell.

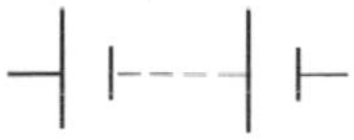

   Fig. 2

   3. Primary cell supplies electrical energy.

   Fig. 3

4. Triac, a three terminal bidirectional device which contains back to back thyristers.

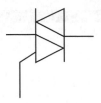

Fig. 4

5. A polarized capacitor. This must be connected the correct way round or it will be damaged.

Fig. 5

6. A variable capacitor.

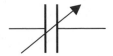

Fig. 6

7. A preset variable capacitor (trimmer).

Fig. 7

8. A d.c. relay can be used for circuit control.

Fig. 8

9. Diac. A two terminal device which is back to back thyristers. This device is triggered on both halves of each cycle.

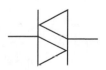

Fig. 9

10. Light-sensitive diode.

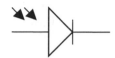

Fig. 10

11. Light-emitting diode (LED). Converts electrical energy to light.

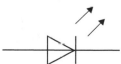

Fig. 11

12. Zener diode. This device acts the same as a diode, but will conduct in the reverse direction a predetermined voltage. It is used for voltage regulation.

Fig. 12

13. Diode will conduct in one direction only.

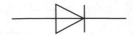

Fig. 13

14. Fuse link.

Fig. 14

15. Iron-cored inductor. A coil of wire which creates a magnetic field when a current is passed through it. Can be used on an a.c. circuit to create a high voltage when the magnetic field collapses or to restrict the flow of current (choke in fluorescent fitting).

Fig. 15

16. Air-cored inductor (as 15).

Fig. 16

17. Inverter. Changes d.c. to a.c. current. Useful for motor control as the frequency can be altered. The waveform is rectangular, fortunately most a.c. motors and fluorescent lamps can accept these waveforms.

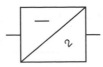

Fig. 17

18. Variable resistor. Potentiometer, 3 contact device used to control voltage.

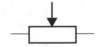

Fig. 18

19. Fixed resistor.

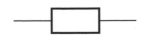

Fig. 19

20. Variable resistor. Rheostat, two terminal device used to control current.

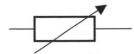

Fig. 20

21. Preset resistor.

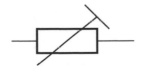

Fig. 21

22. Thermistor. Resistance alters due to heat, a negative coefficient type reduces resistance as it gets hotter, a positive coefficient type increases resistance as it gets hotter.

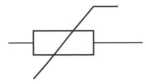

Fig. 22

23. Rectifier. Converts a.c. to d.c. current.

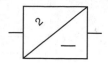

Fig. 23

24. Solenoid valve.

Fig. 24

25. Three-phase star supply.

Fig. 25

26. Three-phase delta supply.

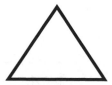

Fig. 26

27. NPN transistor. Amplifies current or can be used with other electronic components to make a switching circuit.

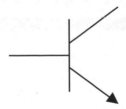

Fig. 27

28. PNP transistor. As 27.

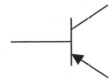

Fig. 28

29. Light-sensitive transistor.

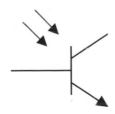

Fig. 29

30. Transformer.

Fig. 30

Glossary

a.c.	Alternating current
Area	Extent of a surface
BS 7671	British standard for electrical wiring regulations
Capacitive reactance	The effect on a current flow due to the reactance of a capacitor
Circle	Perfectly round figure
Circuit breaker	A device installed into a circuit to automatically break a circuit in the event of a fault or overload and which can be reset
Circuit	Assembly of electrical equipment which is supplied from the same origin and protected from overcurrent by a protective device
Circumference	Distance around a circle
Conductor	Material used for carrying current
Coulomb	Quantity of electrons
Correction factor	A factor used to allow for different environmental conditions of installed cables
CSA	Cross-sectional area
Current	Flow of electrons
Cycle	Passage of an a.c. waveform through 360°
Cylinder	Solid or hollow, roller-shaped body
d.c.	Direct current
Dimension	Measurement
Earth fault current	The current which flows between the earth conductor and live conductors in a circuit
Earth fault loop impedance	Resistance of the conductors in which the current will flow in the event of

	an earth fault. This value includes the supply cable, supply transformer and the circuit cable up to the point of the fault
Efficiency	The ratio of output and input power
Energy	The ability to do work
e.m.f.	Electromotive force in volts
Frequency	Number of complete cycles per second of an alternating wave form
Fuse	A device installed in a circuit which melts to break the flow of current in a circuit
Force	Pull of gravity acting on a mass
Hertz	Measurement of frequency
Impedance	Resistance to the flow of current in an a.c. circuit
Impedance triangle	Drawing used to calculate impedance in an a.c. circuit
Internal resistance	Resistance within a cell or cells
Kilogram	unit of mass
kW	True power ($\times 1000$)
kVA	Apparent power ($\times 1000$)
kVAr	Reactive power ($\times 1000$)
Load	Object to be moved
Load	The current drawn by electrical equipment connected to an electrical circuit
Mutual induction	Effect of the magnetic field around a conductor on another conductor
Magnetic flux	Quantity of magnetism measured in Webers
Magnetic flux density	Is the density of flux measured in Webers per metre squared or Tesla
Newton	Pull of gravity (measurement of force)
On-Site Guide	Publication by the IEE containing information on electrical installation
Ohm	Unit of resistance
Overload current	An overcurrent flowing in a circuit which is electrically sound

Percentage efficiency	The ratio of input and output power multiplied by 100
Power	Energy used doing work
Pressure	Continuous force
Primary winding	Winding of transformer which is connected to a supply
Perimeter	Outer edge
Potential difference	Voltage difference between conductive parts
Prospective short circuit current	The maximum current which could flow between live conductors
Prospective fault current	The highest current which could flow in a circuit due to a fault
Protective device	A device inserted into a circuit to protect the cables from overcurrent or fault currents
Resistor	Component which resists the flow of electricity
Resistance	Opposition to the flow of current
Resistivity	Property of a material which affects its ability to conduct
Rectangle	Four-sided figure with right angles
Space factor	Amount of usable space in an enclosure
Secondary winding	Winding of transformer which is connected to a load
Self-Induction	Effect of a magnetic field in a conductor
Series	Connected end to end
Thermoplastic	Cable insulation which becomes soft when heated and remains flexible when cooled down
Transpose	Change order to calculate a value
Triangle	Three-sided object
Thermosetting	Cable insulation which becomes soft when heated and is rigid when cooled down
Transformer	A device which uses electromagnetism to convert a.c. current from one voltage to another

Voltage drop	Amount of voltage lost due to a resistance
Volume	Space occupied by a mass
Watt metre	Instrument used to measure true power
Waveform	The shape of an electrical signal
Work	Energy used moving a load (given in Newton metres or joules)
Phasor	Drawing used to calculate electrical values

Answers

Exercise 1

1. 2768 W   2. 450 000 Ω   3. 37 mA   4. 3300 V
5. 596 Ω   6. 49.378 kW   7. 0.0165 A   8. 132 kV
9. 0.000 001 68 C   10. 0.724 W
11. (a) 0.000 003 6 Ω, (b) 1600 Ω, (c) 85 000 Ω, (d) 0.000 020 6 Ω,
    (e) 0.000 000 68 Ω
12. (a) 1850 W, (b) 0.0185 W, (c) 185 000 W, (d) 0.001 850 W,
    (e) 18.5 W
13. (a) 0.0674 V, (b) 11 000 V, (c) 240 V, (d) 0.009 25 V, (e) 6600 V
14. (a) 0.345 A, (b) 0.000 085 4 A, (c) 0.029 A, (d) 0.0005 A,
    (e) 0.0064 A
15. 139.356 Ω   16. 5040 W
17. (a) 5.3 mA, (b) 18.952 kW, (c) 19.5 MΩ, (d) 6.25 μC,
    (e) 264 kV
18. 5.456 kW   19. 594 250 Ω   20. 0.0213 A
21. 0.000 032 5 C   22. 0.004 35 μF
23. (d)   24. (d)   25. (a)   26. (c)

Exercise 2

1. 45.5 m^3 2. 0.147 m^3 3. (a) 0.375 m^3, 3.25 m^3
                                       (b) 0.0624 m^3, 0.6284 m^2
4. 1.2 m^3 5. 491 mm^3 6. 1313.5 m^3
7. 44.52 m^3 8. 9 m^2 9. 2.755 m   10. 0.291 m^2

Exercise 3

1. 10.5 m^2 2. 0.14 m

3.
Length (m)	6	3	12	8	12
Breadth (m)	2	2	7	4	4

Perimeter (m)	16	10	38	24	32
Area (m²)	12	6	84	32	48

4. 10.5 m^2

5. 19 m^2

6.

Base (m)	0.5	4	1.5	11.25	0.3
Height (m)	0.25	4.5	2.2	3.2	0.12
Area (m²)	0.625	9	1.65	18	0.018

7.

Area (m²)	0.015	0.25×10^{-3}	7.5×10^{-3}	0.000 29	0.0016
Area (m²)	15×10^3	250	7500	290	1.6×10^3

8.

Diameter	0.5 m	0.318 m	0.7927 m	2.76 mm	4 mm
Circumference	1.571 m	1.0 m	0.252 m	8.67 mm	12.57 mm
Area	0.196 m²	0.079 m²	0.5 m²	5.98 mm²	12.57 mm²

9. 0.331 m^2 (575 mm × 575 mm), 9 rivets

10. 0.633 m^2 (660 mm × 958 mm), 2.64 m angle, 80 rivets

11. 19.6 m

12.

No. and diameter (mm) of wires	1/1.13	1/1.78	7/0.85	7/1.35	7/2.14
Nominal c.s.a. (mm²)	1	2.5	4	10	25

13.

Nominal and overall diameter of cable (mm)	2.9	3.8	6.2	7.3	12.0
Nominal overall cross-sectional area (mm)	6.6	11.3	30.2	41.9	113

14. (a) 133 mm^2, (b) 380 mm^2, (c) 660 mm^2

15.

	Permitted number of pvc cables in trunking of size (mm)		
Cable size	50 × 37.5	75 × 50	75 × 75
16 mm²	20	40	60
25 mm²	13	27	40
50 mm²	8	15	22

16. 75 mm × 50 mm or 100 mm × 37.5 mm

17. 8036 mm^2 (about 90 mm × 90 mm) use 100 mm × 100 mm

18. (c)   **19.** (b)   **20.** 75 mm × 75 mm trunking

21. 19 pairs can be added

22. 25 mm conduit, draw in box after second bend

23. (a) 32 mm conduit, (b) adequate room exists but re-calculation of new and existing cable ratings will be necessary

24. (a) 50 mm × 50 mm or 75 mm × 38 mm trunking, (b) 32 mm conduit, (c) difficulty may result when extending from stop-end of 75 mm × 38 mm trunking

Exercise 4

1. 3.21 sec   **2.** 20 sec (2 min)   **3.** 3,870 C

Exercise 5

1. (a) 106 Ω,   (b) 12.5 Ω,   (c) 24 Ω,   (d) 1.965 Ω,
(e) 154.94 Ω, (f) 346.2 Ω,   (g) 59.3 kΩ, (h) 2 290 000 Ω,
(i) 0.0997 Ω, (j) 57 425 μΩ

2. (a) 22 Ω,   (b) 2.35 Ω,   (c) 1.75 Ω, (d) 2.71 Ω,
(e) 1.66 Ω,   (f) 13.42 Ω,   (g) 6.53 Ω, (h) 1805 Ω,
(i) 499 635 μΩ, (j) 0.061 MΩ

3. 3.36 Ω   **4.** 21.1 Ω   **5.** 9   **6.** 533 Ω, 19

7. 133.6 Ω, 30.4 Ω   **8.** 2.76 Ω   **9.** (c)   **10.** (d)

11. (b)   **12.** (b)

Exercise 6

1. (a) 1 Ω,   (b) 1.58 Ω, (c) 3.94 Ω, (d) 1.89 Ω, (e) 2.26 Ω,
(f) 11.7 Ω, (g) 6 Ω,   (h) 5 Ω,   (i) 10 Ω

2. (a) 16 Ω,   (b) 6.67 Ω, (c) 7.2 Ω, (d) 6 Ω, (e) 42 Ω,
(f) 2000 Ω, (g) 300 Ω, (h) 37.5 Ω, (i) 38 Ω, (j) 17.3 Ω

3. (a) 2.13 A, (b) 8.52 A **4.** (a) 4.2 V, (b) 3.53 A, 26.4 A

5. 0.125 Ω   **6.** 25 A, 24.4 A, 21.6 A   **7.** 10.9 Ω

8. 40 A, 30 A   **9.** 20 A   **10.** 0.1 Ω

11. (a) 0.009 23 Ω, (b) 1.2 V, (c) 66.7 A, 50 A, 13.3 A

12. (a) $I_A = I_B = I_D - 2.5$ A, $I_C = 6.67$ A, (b) 9.17 A,
(c) $U_A = 10$ V, $U_B = 12.5$ V, $U_C = 40$ V, $U_D = 17.5$ V,
(d) 9.4 A

13. (d)   **14.** (a)   **15.** 0.365 Ω   **16.** 24 V, 12.4 V

17. (a) 0.0145 Ω, **18.** 141 A, 109 A   **19.** 0.137 Ω
(b) 400 A

20. 971 Ω   **21.** (a) 152 A, 121 A, 227 A, (b) 1.82 V, (c) 1.82 V
22. (b)   **23.** (b)

Exercise 7

1. (a) 3.6 Ω, (b) 5 Ω, (c) 4 A in 9 Ω
resistance, 6 A in 6 Ω resistance, 10 A in 1.4 Ω resistance
2. (a) 2 Ω, (b) 12 V   **3.** 2.25 Ω   **4.** 2.86 Ω
5. (a) 5.43 Ω, (b) 3.68 A   **6.** 26.2 V
7. 0.703 A in 7 Ω resistance, 0.547 A in 9 Ω resistance
8. 11.7 Ω   **9.** 0.174 A, 0.145 A, 50.6 W   **10.** 37.73 Ω
11. 25 .4 Ω   **12.** (a) 20.31 A, (b) 0.23 Ω, (c) 2.64 kW
13. 216.3 V, 213.37 V
14. (a) 5 Ω, (b) $I_A = 3$ A, $I_B = 12$ A, $I_C = 15$ A   **15.** 6 Ω
16. (a) 20 A, (b) 2 Ω, (c) 56 A   **17.** 2.96 V   **18.** 4.25 V
19. 6.96 V   **20.** 0.0306 Ω
21. (a) 1.09 A, 0.78 A, (b) 5.44 V, (c) 1.05 W
22. (a) 4.37 V, (b) 0.955 W, 0.637 W
23. 88.7 V   **24.** 107 V   **25.** 117 V, 136 W, 272 W
26. 2.59 Ω   **27.** (a) 4.4 Ω, (b) 15.9 V, (c) 1.72 Ω
28. (b)   **29.** (d)   **30.** (a)

Exercise 8

1. 1.2 Ω   **2.** 0.222 Ω   **3.** 6 mm^2 **4.** 10 mm^2
5. 43.2 m   **6.** 4.17 V, 6.66 V   **7.** 0.001 48 Ω   **8.** 6 mm
9. 0.623 Ω   **10.** 1.71 V   **11.** 48 m   **12.** 0.071 Ω
13. 47.4 m   **14.** 0.006 Ω   **15.** 25 mm   **16.** 2.5 mm^2
17. (a) 1.187 Ω, (b) 0.297 Ω, (c) 0.178 Ω,
(d) 0.051 Ω, (e) 0.0356 Ω
18. (a) 0.155 Ω, (b) 0.111 Ω, (c) 0.34 Ω
19. (a) 1.32 Ω, (b) 0.528 Ω, (c) 8.8 Ω   **20.** 0.15 Ω
21. 2.63 Ω   **22.** (a)   **23.** (c)   **24.** (a)

Exercise 9

1.

P (watts)	1440	3000	1600	1000	20	1000	2350	1080
I (amperes)	6	12	6.67	150	0.2	5.45	5.1	4.5
U (volts)	240	250	240	6.67	100	220	460	240

2. 30 W   **3.** 4370 W

4. (a) 13.04 A, (b) 6.53 A, (c) 1.96 A, (d) 15.22 A, (e) 30.44 A,
(f) 0.26 A, (g) 0.43 A, (h) 8.7 A, (i) 3.26 A, (j) 0.065 A

5. 108 V   **6.** 203 W   **7.** (a) 18.75 A, (b) 20.45 A   **8.** 748 W

9. 6 A   **10.** 13.04 A   **11.** 633 W

12. 0.2 A, 23 W   **13.** 17.4 A

14. (a) 6.413 kW, (b) 83.7 W, (c) 222 V

15. 29.2 Ω   **16.** (a) 460 V, (b) 7.05 kW, (c) 76.9 W

17. (a) 62.1 W, 70.9 W, (b) 286 W, 250 W

18. 453 W, 315 W   **19.** 40.5 V, 0.78 Ω, 750 W

20. (a) 434 V, (b) 15.6 kW, (c) 216 W, (d) 433 V

21. (c)   **22.** (b)   **23.** (c)   **24.** (a)

Exercise 10

1.

Power (W)	1500	200	1800	1440	1000	2640	100	42.25
Current (A)	10	5	15	12	4.2	11.49	0.42	1.3
Resistance (Ω)	15	8	8	10	56.7	20	567	25

2. 130 W   **3.** 29.4 W   **4.** 5 A   **5.** 1.73 kW

6. 576 Ω   **7.** (a) 6.21 V, (b) 71.4 W   **8.** 530 W

9. 20 A   **10.** 59.4 W   **11.** 125 Ω   **12.** 63.4 W

13. 419 W   **14.** 0.248 W   **15.** 0.8 A

Exercise 11

1. 72 W   **2.** 130 Ω   **3.** 4 Ω   **4.** 120 V

5.

Power (W)	128	100	60	125	768	1800	42.24	36
Voltage (V)	80	240	250	50	240	220	3.5	12
Resistance (Ω)	50	576	1042	20	75	26.9	0.29	4

6. 15 V   **7.** 557 W   **8.** 52.9 Ω   **9.** 170 W   **10.** 161 Ω

11. (a) 28.8 Ω, (b) 19.2 Ω, (c) 16.5 Ω, (d) 128 Ω, (e) 960 Ω,
(f) 8.23 Ω, (g) 576 Ω, (h) 38.4 Ω, (i) 76.8 Ω, (j) 14.4 Ω,

12. 79.1 V   **13.** 4.19 A   **14.** (a) 0.149 A, (b) 29.8 W

15. 115 Ω, 28.8 Ω, 500 W, 2000 W   **16.** 19 A, 126 W, 81 A, 538 W

17. 750 W, 3000 W

18.

Power (W)	25.6	250	360	400	600	960
Current (A)	0.8	2.5	3	3.15	3.87	4.9
Resistance (Ω)	40	40	40	40.3	40	40

(a) 550 W approx., (b) 4.4 A approx.

19.

Power (W)	3000	2000	1661	750	420	180
Voltage (V)	240	200	180	120	89.6	58.8
Resistance (Ω)	19.2	20	19.5	19.2	19.1	19.2

(a) 175 V, (b) 3200 W approx.

22. (d)   **23.** (b)   **24.** (d)   **25.** (c)

Exercise 12

1. 210 Nm   **2.** 208 N

3.

F (N)	85	41.7	0.25	6.5	182
I (m)	0.35	1.2	0.6	0.125	2.75
M (Nm)	29.8	50	0.15	0.813	500

4. 66 N   **5.** 0.3 m   **6.** 37.5 J   **7.** 20 m

8. 20 000 J   **9.** 306 400 J   **10.** (a) 11.5 mJ, (b) 2.5 mW

11. 144 W

12.

Distance between effort and pivot (m)	1	1.5	1.25	0.6	1.8
Distance between load and pivot (m)	0.125	0.3	0.15	0.1	0.2
Load (kgf)	160	200	416.5	390	225
Effort (kgf)	20	40	50	65	25
Force ratio	8	5	8.33	6	9

13.

Load to be raised (kgf)	250	320	420	180	500
Effort required (kgf)	125	150	75	80	214.3
Vertical height (m)	3	4	0.89	2.4	1.8
Length of inclined plane (m)	6	8.53	5	5.4	4.2

14. (a) 2.55 kgf, (b) 2160 kgf, (c) 0.424 m

15.

Radius of wheel R (cm)	25	16	20	17.5	30
Radius of axle r (cm)	8	6.5	6	8.5	8.5
Load (kgf)	200	185	255	150	175
Effort (kgf)	64	75	76.5	72.9	49.6

16. (b) 137.5 kgf, (c) 91.7 kgf, (d) 68.8 kgf

17. (b) 170 kgf, (c) 255 kgf, (d) 340 kgf

18. 70.3 kgf   **19.** (a) 184 W, (b) 245 W, (c) 882×10^3 J

20. 88.9%   **21.** (a) 6.1 A, (b) 38.1 A, (c) 13.5 A, (d) 21.3 A, (e) 58.3 A

22. 1.19A   **23.** 12.58 p

24. (a) 4.57 A, (b) 15.35 p, (c) 1.36 mm^2 (1.5 mm^2)

25. 13.2 kW (15 kW), 28.7 A

26. (a) 3.74 V   **27.** 2.892 kW, 14.8 A   **28.** (c)

29. (b)   **30** (c)

Exercise 13

1. (a) 1.25 mWb, (b) 0.795

2.

Wb	0.025	0.035	0.059	0.74	0.85
mWb	25	35	59.5	740	850
Wb	25 000	35 000	59 500	740 000	850 000

3. 0.0688 Wb   **4.** 0.792 T   **5.** 0.943 mWb   **6.** 1.919 N

7. 0.778 T

8.

Flux density (T)	0.95	0.296	1.2	0.56	0.706
Conductor length (m)	0.035	0.12	0.3	0.071	0.5
Current (A)	1.5	4.5	33.3	0.5	85
Force (N)	0.05	0.16	12	0.02	30

13. 0.08 V   **14.** 2.56 m/s   **15.** 180 V   **16.** 102 V

17. 6.5 V   **18.** 0.15 V   **19.** 100 H   **20.** 1.75 H

21. 50 V   **22.** 66.7 A/s   **23.** 30 H   **24.** 80 V

25. (a)   **26.** (c)   **27.** (b)   **28.** (c)

29. (a)   **30.** (c)   **31.** (d)

Exercise 14

1. 3.96 V   **2.** 36.05 V   **3.** 59.1 A   **4.** 81.51 A

5. (a) 2.764 V (b) 227.23 V   **6.** (a) 45 A (b) 3.97 (c) 16 mm^2

Exercise 15

1.

U (volts)	10	20	30	40	50
I (amperes)	1	2	3	4	5
R (ohms)	10	10	10	10	10

2. 36 V

3.

U (volts)	240	240	240	240	240
I (amperes)	12	6	4	3	2.4
R (ohms)	20	40	60	80	100

4. 3 A, 23 Ω

5.

U (volts)	100	100	96	56	96	132	84	144	121	63
I (amperes)	10	10	12	7	8	12	7	12	11	9
R (ohms)	10	10	8	8	12	11	12	12	11	7

6.

I (amperes)	100	10	10	0.1	0.1	0.1	100	0.001	0.1	200
R (ohms)	0.1	1000	0.1	1000	0.1	1000	0.1	2000	2000	0.01
U (volts)	10	100	1	10	0.01	100	10	20	200	2

7.

R (ohms)	480	14	500	16	110	0.07	12	500	0.75	15
I (amperes)	0.5	15	0.05	6	1.2	0.9	0.7	0.2	8	8
U (volts)	240	210	25	96	132	0.063	8.4	100	6	120

8. 2.041 V
9. (a) 0.33 Ω, (b) 0.17 Ω, (c) 0.12 Ω (d) 0.7 Ω, (e) 0.045 Ω
10. 11 A
11. Section SA 2.916 V, section AB 4.253 V, section AC 2.322 V
volts at A = 47.08 V, volts at B = 42.83 V, volts at C = 40.51 V

12.

Rated current (A)	5	15	30	60	100
Minimum fusing current (A)	7	21	42	84	140

13.

Rated current (A)	5	15	30	60	100
Minimum fusing current (A)	6	18	36	72	120

14. (a) 0 (fuse will blow), (b) 216.48 V, (c) 15.33 V, (d) 18.4 V,
(e) 184 V, (f) 191.6 V
15. (b)   16. (c)   17. (c)   18. (a)

Exercise 16

1. (a) £12.20, (b) £207.40   2. (a) £1.32, (b) £322.95
3. (a) £2.83, (b) £467   4. £100.11   5. £84.60
6. £268.08   7. £535.04   8. (a) £1939.87, (b) £2230.85
9. £188.25   10. £672.25   11. (a) £297.90, (b) £349.98
12. (a) (i) £427.73, (ii) 74.85, (b) (i) £531.72, (ii) £93.05,
(c) (i) £1221.66, (ii) 213.70, (d) (i) £62.87, (ii) £11.00
13. (a) £558.51, (b) £625.81, (c) £762.07, (d) £872.57,
(e) £2787.85
14. (c)   15. (b)

1.

Applied volts	50	60	25	80	45
Capacitance (μF)	0.2	0.3	0.4	0.6	0.8
Charge (μC)	10	18	10	48	36

2. (a) 1.88 μF, (b) 450 μC, (c) 150 V, 90 V, (d) 0.34 J, 0.02 J
3. 8 μF, 2.4 μF   4. 60 μF
5. (a) 3.7 μF, 111 V, 74 V, 56 V, (b) 36 μF, 240 V
6. 30 μF   7. (b)   8. (a)   9. (c)